FOREIGN INVESTMENT AND ECONOMIC DEVELOPMENT IN CHINA: 1979-1996

To my parents and Ping Wang for all their support and love

Foreign Investment and Economic Development in China: 1979-1996

HAISHUN SUN
Deakin University

Ashgate

Aldershot • Brookfield USA • Singapore • Sydney

332.6730951
S95f

Published by
Ashgate Publishing Ltd
Gower House
Croft Road
Aldershot
Hants GU11 3HR
England

Ashgate Publishing Company
Old Post Road
Brookfield
Vermont 05036
USA

British Library Cataloguing in Publication Data
Sun, Haishun
    Foreign investment and economic development in China:
    1979-1996
    1. Investments, Foreign - China 2. China - Economic
    conditions - 1976-
    I. Title
    338.9'51

Library of Congress Catalog Card Number: 97-78324

ISBN 1 84014 354 1

Printed and bound by Athenaeum Press, Ltd.,
Gateshead, Tyne & Wear.

Contents

List of Tables

Acknowledgment

This book is an outgrowth of my Ph.D thesis submitted to the University of Queensland. It is an empirical analysis of the impact of direct foreign investment on the economic development of China over the period 1979 to 1996, which is one of the most important and intriguing facets of the Chinese economy during this period.

I wish to make my major acknowledgment to Dr. Joseph C. H. Chai, my thesis supervisor at the University of Queensland, who has given me not only advice but more importantly encouragement ever since I started this research in 1992. I must also thank Professor Clem Tisdell, Dr. Mohammad Alauddin and other staff members in the Department of Economics, the University of Queensland, for their advice, supports and assistance for this research.

My great appreciation must also go to School of Economics, Faculty of Business and Law, Deakin University for providing academic environment and research facilities, which enable me to complete the final version of the book. In particular, I would like to extend my gratitude to Professor Pasquale M. Sgro and Dr. Phillip Hone for their supports for my work on this book. It is no exaggeration to say that without their supports, this book could never have been completed.

I am also grateful to Professor Ru Xin, the Vice-President of Chinese Academy of Social Sciences, for his constant support and encouragement for this research. My sincere thanks must also go to Professor Allan Hope, Professor Y.Y. Kueh and Professor Peter Groenewegen, who have read the earlier versions of the book and provided a lot of valuable comments and advice on this study. Specifically, I would like to thank Ms. Fran Cozine and Mr. Carl Cozine for their encouragement and support for my study and research over the past several years.

Financial support from the World Bank through a graduate scholarship program and editorial advice by Ms. Kate Hargreave, Mrs. Rachel Hedges and Ms. Anne Keirby from Ashgate Publishing Group are gratefully acknowledged. Finally, I would like to express my deep gratitude to my beloved parents and my wife, Ping Wang, for their support and love.

List of Abbreviations

DFI: direct foreign investment
FIEs: foreign-invested enterprises
GDP: gross domestic product
TVEs: township and village enterprises
MNCs: multinational corporations
SOEs: state-owned enterprises
EJVs: equity joint ventures
CJVs: contractual joint ventures
WFOEs: wholly foreign-owned enterprises

1 Impact of Direct Foreign Investment: Issues and Theories

1.1 The Issue of DFI in the Context of China

China's economic reforms and 'open-door' policy since 1979 have changed the orientation of her industrialization drive and brought dramatic changes in the economy. China achieved a remarkably high average growth rate of 9.9 percent per annum during the period 1979 to 1996, which places her among the most dynamic economies in the world. In large part, this growth is attributed to an increasing 'opening up' of China to the outside world.

China's merchandise exports grew at an annual average rate of 15.8 percent in the same period (SSB, 1997), the highest growth rate in the Asia-Pacific Region. More importantly, direct foreign investment (DFI) witnessed exceptional growth and exerted a far-reaching impact on the Chinese economy. The realized DFI grew by 38.7 percent per year from 1981 to 1996, with the cumulative realized DFI reaching US$174.9 billion by the end of 1996. China has effectively become the largest recipient of DFI in the developing world.

In terms of population, gross domestic product (GDP) and geographical size, China is the largest developing country undergoing economic transition from a socialist central planning system to a market economy. Its experience in using foreign investment in economic development provides valuable lessons for both developing countries and others - specifically economies in transition such as Eastern Europe, the former Soviet Union, and Asia. The most important issue for policy makers in the host country is to provide an appropriate economic environment and policies for generating maximum benefit from DFI.

Debate amongst economists and policy makers about the role of DFI in the host economy falls into two schools. The first regards DFI as a dynamic force in the host country. They believe that DFI makes a positive contribution to the host country's economic growth through an increasing capital supply, technology transfer, training and productivity gains. The second school considers DFI as a tool of international exploitation by multinational corporations (MNCs). They view investment by MNCs in a developing country as a leading ultimately to economic dependence by the host country on these MNCs, which undermines the host country's economic authority. This is because MNCs are motivated to invest in order to exploit the natural resources and cheap labor available in the host country. They see the benefits produced by these investments as mainly accruing to MNCs through transfer pricing for imports and exports. As a result, the economic welfare of the host country is not improved.

Over the past three decades, various countries have experienced flows of foreign investment with differing outcomes. The four Newly Industrialized Economies (NIEs) in East Asia (Hong Kong, Taiwan, Singapore and South Korea) are successful examples of using foreign investment to promote export-oriented economic development. However, in Latin America and Africa, the positive role of DFI in the host economies has been less evident.

Previous studies of the impact of DFI and the role of multinational corporations in host economies were largely based on the experiences of market economies. Such studies discussed this issue primarily in the context of either developed or developing market economies, especially small open economies such as Asian NIEs. However, these studies have provided little evidence of the impact of DFI on a developing economy of China's size in the process of transition.

This study investigates the impact of DFI on Chinese economic development from 1979 to 1996. It provides not only an in-depth investigation of the issues covered by the earlier studies (such as the impact of DFI on domestic capital formation, exports, employment and economic growth), but also explores other development issues ignored by previous studies. These include the industrial linkage effects of DFI on the domestic sector, comparative analysis of the productive efficiency of foreign-invested enterprises (FIEs) and domestic firms, and the impact of DFI on regional economic disparity and income inequality. The entry modes of MNCs and their transfer pricing practices are also investigated. The findings derived from this study highlight lessons for other developing countries, especially those economies undergoing transition from a centrally-planed system to a market economy.

To investigate the impact of DFI on the economic development of China, it is helpful to consider the relevant theories where the role of DFI in the economic growth of developing countries has been a focus of economic research and policy formulation. While a number of theoretical analyses and empirical studies of the impact of DFI have appeared in the economic literature, to date, no single theory has been able to provide a comprehensive and widely accepted explanation of the role of foreign investment in the economic development of developing countries.

This issue involves not only DFI and multinational corporations but also economic growth and development, and it is necessary to incorporate the theories of DFI and MNCs into economic development theories. This is a complex task as the theories of DFI and MNC are essentially micro-economic analyses of transnational investment activities by MNCs, while the economic growth and development theories explore the macro-conditions or determinants of economic growth and development. This difficulty may partially account for the separation of foreign investment theories from the economic development theories, and explain why no broadly agreed theory on the role of DFI in economic development has been developed to date.

1.2 Theories of Direct Foreign Investment

A number of competing and complementary theories have been proposed to explain the nature, causes and possible socio-economic consequences of the rapid growth of DFI. Among others, these include the neoclassical theory of capital mobility, the industrial organization approach, the transaction cost or internalization theory, Dunning's eclectic theory and Kojima's macroeconomic theory of DFI.

The neoclassical theory of capital movement or foreign investment, is part of the theory of international factor movements. Based on the Hecksher-Ohlin-Samuelson (HOS) model, international movements of factors of production, including foreign investment, are determined by the different proportions of the primary production inputs available in different countries. International capital movement implies a flow of investment funds from countries where capital is relatively abundant to countries where capital is relatively scarce. Effectively, capital moves from countries with low marginal productivity of capital to countries with high marginal productivity of capital (Bos et al, 1974). Such international investment (or capital movement) may benefit both the investing and host countries. The host country may benefit from foreign investment to the extent that the productivity of the investment, as reflected in the income created, exceeds what foreign investors take out of the host country in the form of profit and interest.

However, the neoclassical theory with its assumptions of perfect competition, zero transaction costs and perfect information, fails to fully account for investors' motivations and behavior, and impacts of investment on a real world with imperfect markets and uncertainty. In particular, the neoclassical theory does not distinguish DFI from other forms of capital flow and also fails to explain the two-way capital flows between capital-abundant countries, for instance, DFI between the US and Japan or between the US and Britain. The neoclassical failure to explain and predict how, where and why DFI occurs and to highlight the social and economic consequences of DFI has led to the development of new explanations of international investment. These theories include the industrial organization theory, location theory, transaction cost or internalization theory, Dunning's eclectic theory and Kojima's macroeconomic theory.

Industrial organization theory was pioneered by Stephen Hymer in 1960. He pointed out that the movement of capital associated with DFI is not a response to higher interest rates in 'host' countries but takes place in order to finance international operations. His explanation of why firms move abroad and establish international production is based on a theory of the firm and industrial organization. It is primarily concerned with the characteristics of multinational corporations and the market structures in which they operate.

According to this theory, market structures and competition conditions are important determinants of the types of firms which engage in DFI. The theory uses firm-specific advantages, such as a firm's market position, to explain MNC's international investment. These firm-specific advantages include patents, superior knowledge, product differentiation, expertise in organizational and management skills, and access to overseas markets and to credit. The advantages that certain firms have over competitors in the home country can be extended into foreign markets through international direct investment. Hence, DFI is regarded as a strategy by which oligopolistic MNCs seek to close out market competition by the erection of barriers to entry and is related to firm-specific advantages (McClintock, 1988 and Yamin, 1991).

In contrast, *location theory* places emphasis on country-specific characteristics. It explains DFI activities in terms of relative economic conditions in source and host countries, and considers locations in which DFI activities operate. This approach includes two subdivisions - 'input-oriented' and 'output-oriented.' Input-oriented factors are associated with supply variables such as relative costs of inputs including labor, raw materials, energy and capital. Output-oriented factors focus primarily on the determinants of market demand (Santiago, 1987, p.318). These include the population size, income per capita, and the openness of the markets in host countries. The country-specific factors not only determine where MNCs make direct investment, but can also be utilized to account for the different types of DFI such as domestic-market-oriented investment and export-oriented investment.

The third school of DFI theories is referred to variously as the *transaction cost or internalization approach*. Major advocates in this school are Vernon (1966), Caves (1982), Rugman (1981, 1986), Buckley and Casson (1985), and Hennart (1992). This approach interprets the DFI activities of MNCs as a response to market imperfection which causes increased transaction costs. Applying the transaction cost theory proposed by Williamson (1973), this approach explains why the firm by internalizing economic activities may provide more efficient outcomes than the market in minimizing transaction costs. Through direct foreign investment, two types of market imperfection may be internalized. The first one is structural (or institutional) market imperfection associated with regulatory aspects such as tariffs or subsidies, foreign exchange controls, import quotas, income taxes, restrictions on profit repatriation and other restraints. MNCs tend to internalize this type of market imperfection for a rent-seeking purpose.

Market imperfection also relates to market transaction costs, especially for an intangible asset transaction such as technology transfer. In order to exploit new advantages and to exert full control, MNCs prefer DFI rather than trade or licensing the use of their firm-specific intangible assets. This allows MNCs to maintain market share or market power and to maximize their benefit. In essence, the transaction cost theory or internalization theory is a model of private welfare maximization based on MNC's operation (McClintock, 1988). However, it allows little room for social welfare considerations either on a national or global scale, and therefore fails to account for the macroeconomic impact of DFI.

The *eclectic theory of international production* developed by Dunning (1977, 1981, 1988), combines the industrial organization approach with both the location theory and internalization theory to explain DFI and international production activities. Dunning sees three conditions as indispensable for DFI. First, a firm must possess net ownership advantages over rival firms in the host country's market. Second, it must be more profitable for the firm to maintain these advantages internally, rather than to sell or lease them to foreign firms. Finally, the firm must believe that its advantages can be better exploited by using location-specific factors (such as labor and market) in the host countries than by simply exporting to foreign markets (Santiago, 1987). Since the eclectic theory of international production encompasses complementary aspects of each of the industrial organization approach, internalization theory and location theory, it provides a more comprehensive explanation of the nature and characteristics of DFI initiated by a MNC.

The theories reviewed above are essentially the theories of a particular type of firm, viz. multinational corporations, and provide micro-economic analyses of the nature, causes and types of DFI and of MNCs' transnational business behavior. One serious omission in these studies of DFI and the operation of MNCs is the macroeconomic analysis. As a consequence, they fail to explain international differences in DFI by MNCs from countries with different cultural backgrounds and macroeconomic structures. In addition, these theories fall far short of providing a framework for analysis of the macroeconomic impact of DFI in the host country.

Kojima (1973, 1975 and 1982) has proposed a macroeconomic theory of direct foreign investment within the framework of relative factor endowments. He identifies two different types of DFI: trade-oriented (Japanese type) and anti-trade-oriented (American type) based on comparative advantages and industrial structures. He proposes that DFI from a comparatively disadvantaged industry in the investing country (which is potentially a comparatively advantaged industry in the host country) will harmoniously promote an upgrading of industrial structure on both sides and thus will accelerate trade between the two countries. Kojima argues that comparative profitability in trade-oriented DFI conform to the direction of potential comparative costs and therefore complement each other. He claims that Japanese investments in developing countries of Asia are largely in labor-intensive and resource-based industries, two areas in which the host countries have advantages over Japan. These investments complement Japan's comparative advantage position, thus creating more trade between the investing and receiving countries.

By comparison, American direct foreign investment does not conform to this comparative profitability formula. The US investments abroad are concentrated in capital-intensive and high technology industries in which it has comparative advantages. Kojima argues that DFI by large and oligopolistic firms in these industries is 'anti-trade-oriented' and works against the structure of comparative advantage. This is because these new industries set up foreign subsidiaries, cutting off their own advantages and leading to trade-substitution effects.

Although Kojima's macroeconomic theory of DFI provides a basis for the analysis of the relationship between DFI and trade, it provides little insight for analysis of the impacts of DFI on other macroeconomic variables for both investing and host countries. Kojima also fails to build a plausible microeconomic basis for the macroeconomic theory of DFI (Lee, 1984). In practice, a distinction in type of DFI (so-called trade-oriented and anti-trade-oriented) between Japan and the US does not always exist either at the firm or national levels. The two types of DFI could co-exist in one country, even in one industry. DFI from one country may be distributed in a variety of industries, including advantaged and disadvantaged industries. Therefore, Kojima's classification of two types of DFI makes his macroeconomic approach less valid for assessing the economic impact of DFI in an empirical sense.

An investigation of the macroeconomic impacts of DFI in host countries must study how and to what extent DFI affects the conditions and determinants of economic growth and development. This suggests that the role of DFI in the host economy can be approached within the theoretical framework of economic development. During the past two decades, a number of theoretical and empirical studies have been published. These studies discuss the role of DFI in the economic development of the host countries from different perspectives, and tend to either complement or contradict each other. Accordingly, they suggest that DFI may positively or negatively affect the host country's economy. As Dunning (1988) argues, the benefits to be reaped from DFI critically depend on the type and nature of the investment, the economic conditions and characteristics of the host country, and the macroeconomic and organizational strategies and polices pursued by host country governments.

1.3 Economic Impact of DFI: Supply-Side View

In theory, the impact of DFI on the host economy can be realized in two ways. First, DFI may affect the supply of productive resources including financial capital, equipment and machinery, technology, management expertise and labor training. Second, it can influence the aggregate demand of the host country through initial investment demand and subsequent input demand.

Both the classical and neoclassical economic theories explain economic growth and development in terms of the stock of productive resources available for an economy and the utilization of these resources. The productive resources include capital, labor, technology, management skills and natural resources. According to Ricardo's classical theory of growth, an increase in capital and labor would result in growth of output. In the Harrod-Domar Model of growth, the change in capital stock (investment) and incremental capital-output ratio (ICOR) determine the growth of national income. For a given ICOR, an increase in investment will lead to an increase in income (output). Accordingly, for a given amount of capital, the income is determined by marginal capital productivity (the inverse of ICOR).

In Solow's neo-classical model, economic growth is not only determined by the stock of capital and labor but also by the capital-labor ratio. If capital increases faster than the increase in labor (termed capital deepening), the capital-labor ratio will increase and result in a growth of labor productivity. The modern theories of economic growth extend the analysis of factors contributing to economic growth, with technology and exports also being included in economic growth models. Technological progress, capital deepening, export expansion, and rational management and development strategies, are believed to be critical factors influencing economic growth.

Many development economists argue in the context of developing countries that economic development is restrained by the shortage of capital (both financial and physical), technology, skilled labor and management expertise, and foreign exchange. The shortage of these productive factors, as the 'two-gap' or 'four-gap' models point out, cause the bottlenecks in economic development of developing countries. Removing or alleviating these gaps or bottlenecks, it is argued, is the key for these countries to achieve economic growth and modernization.

Based on the economic growth and development theories, economists, including Ahiakpor (1990), Nuñez, (1990), Hill and Johns (1991), Tu (1990), Lim and Fong (1991), Chen (1979, 1993) and Todaro (1994), propose that DFI may positively affect the economic growth of developing countries through the following channels. First, DFI may positively contribute to the capital formation of the host country. DFI, as a type of foreign capital inflow, represents an addition to the domestic savings of the host country. All other things being equal, this will augment the financial resources available for the domestic investment of the host country. Moreover, DFI may bring advanced equipment and machinery to the developing host country or finance the importation of capital goods that cannot be produced in the host country, thereby contributing to its capital formation.

However, the relationship between DFI and domestic investment by the host country is inconclusive. As Areskoug (1976) argues, it could stimulate, supplement or displace indigenous investment financed by domestic savings. If DFI improves infrastructure in the host country and creates good investment conditions or opportunities, it tends to promote domestically-financed investment. Furthermore, the addition of foreign capital may also relieve pressure on the rate of interest charged in capital markets in the host country, and provide an incentive for domestic investment (Ahiakpor, 1990).

Nevertheless, DFI may displace indigenous investment in the host country. If DFI is financed from the local financial market and results in a higher interest rate, it may crowd out domestic investment (Jansen, 1995). The crowding out may also occur in factor and commodity markets. If foreign investors compete with local firms for use of scarce resources, such as import licenses, skilled manpower and credit facilities, the supply of these scarce resources will decrease for local firms. Likewise, if foreign investors tend to sell their products in the domestic markets and gain an increasing market share, local firms could be crowded out from the industry or at least have to shrink their production. Besides, foreign investors may foreclose investment opportunities for local investors based on their technology advantage and market power. Therefore, the net impact of DFI on capital formation in the host country depends upon its effect on the domestically-financed investment.

DFI may promote productivity of the domestic sector of the host country through technology transfer and the training of local labor, technicians and management personnel. Many economists agree that DFI-induced technology progress, transfer and diffusion are the most important contributions of DFI to the economy of the host country. It is widely believed that the new forms of DFI, especially joint ventures, facilitate the transfer and diffusion of technology in the host country.

In addition, through the forward linkage effect, foreign-invested enterprises (FIEs) supply equipment, machinery and other intermediate products to domestic firms. As the availability of these inputs increases, the production of domestic firms can be stimulated. In addition, the products made by FIEs may also substitute for imported products, thus helping the host country to alleviate reliance on imports and thereby to reduce any trade deficit.

Finally, FIEs are seen to contribute to the host Government's tax revenue. However, the net contribution depends on whether the tax revenue paid by FIEs is larger than the expenditure by the host government for establishing and improving infrastructure for DFI. In this regard, transfer pricing manipulated by MNCs in order to avoid tax is an important factor affecting the host country's tax revenue.

1.4 Economic Impact of DFI: Demand-Side View

In practice, economic growth in a country depends not only on its productive capacity, but also on the extent to which that productive capacity is actually utilized, together with the strength of demand (Sundrum, 1990). An increase in any component of aggregate demand will lead to a rise of the GDP and income level. DFI may contribute to the economic growth of the host country through positively affecting aggregate demand.

In addition to the initial investment demand, subsequent demand by foreign-invested enterprises for inputs of production is even more important. When MNCs make direct investment and establish production subsidiaries in the host country, they need to employ local labor and management personnel and pay them wages and salaries. The employment creation by DFI is important for many developing host countries where the rates of unemployment and underemployment are high. It provides not only income to employees and thus additional savings to the host country, but also helps improve labor productivity of traditional sectors (such as agriculture) by absorbing underemployed or surplus labor from these sectors.

Another important demand-side impact of DFI is its backward linkage effect on the domestic sectors of the host country. Through buying locally made materials and intermediate products, foreign-invested enterprises can create additional demand for products made by local firms. The initial DFI-generated demand will induce multi-rounds of subsequent demand through industrial backward linkage effects. Domestic suppliers will be stimulated to produce more output. As a result, the growth of the entire economy will be encouraged by the increased aggregate demand initiated by FIEs' local purchases.

Furthermore, DFI, especially export-oriented DFI, promotes the exports of the host country. Taking advantage of an abundant and cheap labor resource in the host country, together with their own marketing channels and expertise, foreign-invested enterprises are able to expand export of their products. In addition, by using locally-made materials and intermediate inputs, FIEs tend to promote exports from domestic firms as well. Export expansion, as an important indicator of economic competitiveness, directly stimulates economic growth and improves the industrial efficiency of the host country.

In practice the impacts of DFI on the supply and demand sides are intermingled rather than separate. In terms of impact scope, the impact of DFI on the host economy can be classified into macroeconomic and microeconomic impacts. The former refers to the impact of DFI on the macroeconomic variables, such as GDP growth, total fixed investment, employment, exports and imports, aggregate consumption, government expenditure and tax revenue. The latter concerns the impact of DFI on the economic behavior of individual units including firms and family. For instance, the influences of DFI on the labor productivity and technical and management efficiency of a domestic firm are microeconomic impacts of DFI.

While some impacts of DFI can be quantitatively measured, others cannot be directly measured. For instance, the effects of DFI on GDP growth, capital formation, employment, exports and government tax revenue are measurable, whereas the effects on technology transfer and diffusion efficiency, environmental pollution, access to foreign markets and demonstration effects are difficult to quantify.

There are various reasons for the difficulty in measuring these impacts. Some economic variables are affected simultaneously by multiple factors, including political, cultural and economic factors, and it is difficult to separate one factor's effect from that of others. Income distribution, industrial structure change, environmental pollution and inter-regional economic disparity fall into this group. Moreover, with some impacts such as transfer pricing, it is difficult to acquire reliable and sufficient data although this issue is theoretically clear and quantitatively measurable.

The current study focuses on measurable impacts of DFI in China, using the latest data. These include the effects of DFI on domestic capital formation, economic growth, employment, industrial production, exports and foreign trade balance, industrial linkage effects, the productive efficiency of foreign and domestic firms and regional economic disparity. In addition, other issues such as the entry modes of multinational corporations and transfer pricing are also explored.

2 DFI in China: Development and Characteristics

2.1 Introduction

Direct foreign investment (DFI) in China has increased rapidly since China began pursuing an 'open-door' policy in 1979. As of August 1997, China had approved 266,900 foreign-invested enterprises (FIEs), with a realized investment value of US$204.4 billion (MOFTEC 1997). China is now the largest recipient of DFI in the developing world. In 1995, it absorbed some 40 per cent of the total of US$100 billion invested in developing countries (UNCTAD 1996, p. 50), and was second only to the United Sates as the largest recipient country of DFI in the world.

As DFI flows have increased, their impact on the Chinese economy has become increasingly important. To facilitate an investigation of the impact of DFI on China's economic development and transition in the past 18 years, this chapter presents an analysis of the development and characteristics of DFI in China since 1979, with focus on the policy evolution and a comparative analysis of the investment patterns of major investing countries. It also assesses the factors accounting for the regional distribution of DFI. Finally, an analysis of the industrial composition of DFI by major investors, and the reasons for differences in DFI composition between major investing countries is provided.

2.2 The Three Phases of DFI Development in China: Policy Dimension

While direct foreign investment has experienced a rapid growth since 1979, the growth has been subject to considerable fluctuations at different times. In large part, the fluctuations of DFI reflect adjustments in the Chinese government's economic policies and resulting changes in the investment environment. In general, when China has evidenced a sound macroeconomic performance and economic liberalization is to the forefront, DFI tends to grow rapidly. On the contrary, poor economic performance (low growth and high inflation), has tended to dampen economic reforms and halt the liberalization process. At these times, foreign investment declines

The development of DFI in China has undergone three phases during the last 18 years, reflecting changes in economic conditions, investment environments and government policies.

2.2.1 Phase One: 1979-1985

In mid-1979 the Chinese government promulgated the Joint Venture Law, which was the first formal statement of commitment to foreign investment by the government. It established the principles and procedures for foreign investment (Pearson, 1991). Under this law, foreign investments are permitted and encouraged in selected fields. As an important step to using foreign investment, the Chinese government set up four Special Economic Zones (SEZs) where preferential economic policies were pursued to utilize foreign investment. Following the promulgation of the Joint Venture Law and establishment of SEZs, foreign firms flocked to China to explore the new business opportunities. By the end of 1982, 922 foreign invested projects had been approved with pledged investments of US$4,608 million, of which US$1,168 million was realised (see Table 2.1).

However, the regulatory environment for foreign investment was still quite restrictive, with restrictions on foreign equity share in joint ventures (generally less than 50 percent), and industrial scope (DFI was not permitted in finance and banking, transportation, telecommunication & post, and retail). There were also restrictions on FIEs' access to the domestic markets as well as their use of local labor and land. As a result, DFI in China grew sluggishly, especially during the first four years.

In 1983, the Chinese government put the Joint Venture Implementing Regulations into effect in order to improve the investment environment. The Implementing Regulation legally clarified the status of joint ventures and provided greater detail about China's policy on important issues such as profit repatriation, technology transfer, and foreign exchange. In April 1984, the Chinese government announced that 14 coastal cities would open to foreign investment, expanding the open-door policy from SEZs to other coastal regions. In addition, investment protection treaties were signed between China and a number of investing countries.

As a result of improvements in the legal and operational environments, DFI increased rapidly. The pledged value of DFI grew at 53 percent and 124 percent in 1984 and 1985 respectively, while the realised DFI grew at 98 percent and 32 percent respectively (see Table 2.1). Some well known multinational corporations (e.g. IBM, American Automobile Corporation, Western Oil Corporation, and etc.) made major DFI commitments in the 1984-85 boom. There were two features of the boom. First, as more hotel joint ventures were approved, 'non-productive' (services) investment projects dominated DFI in China. Second, the size distribution of investment prjects was highly skewed (Pomfret, 1991). At the one extrem, many joint ventures were of small capital size, especially those established by Hong Kong firms in service and manufacturing industries, with most of these being capitalized at less than one million dollars. At the other extreme, some projects were large, with capital exceeding several million dollars. These larger projects were typically investments in hotels, oil exploration, automobile and machinery industries.

A sharp decline in the growth rates of both pledged and realised DFI followed the 1984-1985 boom. The decline started from the second half of 1985 and continued until mid-1986. As shown in Figure 2.2, the growth rate of realised DFI declined from 98 percent in 1984 to 32 percent in 1985, and further to 13 percent in 1986. Even more striking was the negative growth of pledged DFI, a leading indicator of foreign investment. The growth rate of the pledged DFI became negative in 1986, a 52 percent decline compared to the previous year (see Table 2.1 and Figure 2.2).

Although a number of factors could account for this sharp drop in DFI, the main reasons related to the economic problems being experienced by China and a growing wariness on the part of foreign investors about the worsening investment environment. The rapid economic growth in 1984 and 1985 gave rise to high inflation and a growing trade deficit in China. This situation led the Chinese government to severely curtail domestic spending of foreign exchange and to tighten approval of so called 'non-productive' DFI projects. These measures had immediate effects on joint ventures. Many joint ventures faced difficulties such as foreign exchange shortage, and increased operating costs. These were exacerbated by limited domestic market access, a low labor productivity and excessive government bureaucracy. Under these circumstances, foreign investors' confidence was frustrated, which resulted in a substantial drop in investment in China. The downturn and depression in pledged and realized DFI lasted almost one year from the last quarter of 1985 to the third quarter of 1986.

2.2.2   Phase Two: 1986-1989

The Chinese government responded to the sharp fall in pledged DFI by publishing the 'Provisions for the Encouragement of Foreign Investment' in October 1986. This marked the beginning of the second phase of DFI development in China. The 1986 Provisions (so called '22 Articles') were followed by a set of central regulations to implement them and by a flurry of provincial and municipal-level regulations (Pearson, 1991). These Provisions and regulations not only clarified the legal environment for FIEs, but also provided the solutions to some major problems. For example, foreign exchange imbalance in FIEs was solved by establishing swap markets between enterprises. In addition, joint ventures were offered tax benefits and greater autonomy for their business management.

A gradual simplification of the joint venture approval process with greater autonomy in decision making at the local level occured after 1986. In addition, provinces and cities provided their own sets of investment incentives in addition to the incentives granted by the central government. These incentives included exemption of local tax, lower land use fees and lower charges for using public utilities. Furthermore, to promote foreign investment, almost all open coastal cities set up the Economic and Technical Development Zones (ETDZs) designed for high technology industrial projects. In the ETDZs extra tax breaks were offered, in addition to lower land-use fees and other incentives. This further enhanced the investment environment in these coastal areas.

In response to the improved investment environment and the preferential treatments, DFI recovered quickly after the autumn of 1986, and maintained a high growth rate in 1987, 1988 and the first half of 1989. The pledged DFI increased by 31 percent in 1987, and by a further 43 percent in 1988. The realised DFI increased by 24 percent and 38 percent respectively in 1987 and 1988 (see Table 2.1). Over the same period the realised DFI in joint ventures grew by 146 percent. In contrast to the prior-1986 structure, 85 percent of DFI projects involved in the manufacturing industries in 1988, and many of these projects were aimed at China's domestic market.

This new foreign investment boom continued until mid-1989 ending with a sharp decline in DFI projects. A worsening economic and political climate was largely to blame. In order to cool down the over-heated economy and to reduce the two-digit inflation, the Chinese government introduced the 'austerity program' in late 1988. This slowed economic growth in 1989 and squeezed the domestic credit and dampened market demand. On the one hand, FIEs geared towards the domestic market found it very difficult to sell their products due to the tightened local loans and the stagnant market demand. On the other hand, the squeezed credit also limited the ability of prospective local partners in joint ventures to raise capital. Many joint ventures ran into financial and marketing difficulties. Some joint ventures had to be halted at the negotiation stage or changed to wholly foreign-owned enterprises.

Post-May and June of 1989 also witnessed a crackdown of the democracy movement by the Chinese government. The reform process halted and a change in policy orientation was apparent. This worsened the investment environment of China and reduced the confidence of foreign investors in the stability of Chinese policies. As a result, the number of joint ventures approved in 1989 decreased by 692 compared to 1988. The growth rate of the pledged DFI value also fell sharply to 5.7 percent in 1989 while realised DFI increased by only 6.2 percent (see Table 2.1). There was an absolute decline in DFI from countries other than Hong Kong and Taiwan in 1989. This situation marked the end of the second boom of DFI which began in late 1986, and indicated that China was entering a new phase of foreign investment development.

2.2.3   Phase three: 1990-1996

In recognition of the negative reaction of foreign investors to the worsened investment environment, the Chinese government issued the Amendments to the Joint Venture Law in April 1990. These amendments codified several rules designed to encourage investment. Particularly important was the elimination of the duration limit applied to some ventures, and granting of permission for foreigners to act as joint venture board chairmen (Robertson and Chen, 1990).

In April 1991, the Income Tax law for Enterprises with Foreign Investment and Foreign Enterprises (The Unified Income Tax Law) was passed by the People's Congress of China. This tax law, as the most significant tax legislation for a decade, replaced both the foreign enterprise and joint venture income laws, which had governed foreign operations in China since the early 1980s. It standardized the income tax rates for different forms of FIEs, and eliminated the prior tax discrimination against the wholly foreign-owned enterprises (WFOEs), which had previously been subject to unfavourable treatment. In addition, the Chinese government concluded the austerity program in early 1991 and replaced it with a promotion for investment through loosening the control over the domestic loans and opening up the domestic market to foreign investors.

The improved investment environment and expansion of domestic credit facilitated the recovery and growth of foreign investment in China after 1990. The pledged DFI grew 18 percent and 82 percent in 1990 and 1991 respectively. More impressively, the pledged DFI in EJVs increased 125 percent in 1991 compared to the previous year. The new upward trend of DFI was also confirmed by a 25 percent increase in the realised DFI (see Table 2.1 and Figure 2.2).

In 1992, the growth of foreign investment was even more remarkable. Following Deng Xiaoping's 'grand tour' of the south early that year, the central government adjusted the economic policy in order to speed up economic reform and to further open the economy to foreign investment. Consequently, the policy on DFI became more favourable. The new policy permits foreign companies to invest in fields such as retailing, real estate, trading, transport, finance and banking, and also permits them to establish stock-holding companies and secondary business to foreign affiliates (Hirano, 1993).

In September 1992, the Chinese government announced the adoption of the 'socialist market economy' strategy and began to build a legal framework to standardise market operation. Regulations covering corporation law, bankruptcy law, individual income law and stock trading law, and some other commercial regulations have been passed since 1993. In order to create a favourable business environment and to facilitate the functioning of market mechanisms, the government has taken a number of reform measures in the recent three years, such as putting into effect the new tax system, uniting the 'two-track' system of foreign exchange rates (officially regulated rates and market rates) and allowing the Chinese currency to be convertable for transactions in current account. In addition, privatisation of state-owned enterprises by selling their shares to the public, and lowering tariff for imports have been important measures further liberalizing the Chinese economy and aiding the emergence of high economic growth and favourable business environment.

Table 2.1 Pledged and Arrived Direct Foreign Investment in China 1979-1996

Year	Number of Projects	Pledged Value (US$ mil.)	Growth Rate %	Average Size of projects (US$ mil.)	Realized Value[1] (US$ mil.)	Growth Rate %	Realization Ratio[2] %
1979-81	783	4078	N/A	5.21	738	N/A	18.1
1982	139	530	N/A	3.81	430	N/A	81.1
1983	470	1732	226.8	3.69	635	47.7	36.7
1984	1856	2651	53.1	1.43	1258	97.8	47.5
1985	3078	5932	123.8	1.93	1661	31.9	28.1
1986	1498	2834	-52.2	1.89	1874	12.8	66.1
1987	2233	3709	30.9	1.66	2314	23.5	62.4
1988	5945	5297	42.8	0.89	3194	38.1	60.3
1989	5779	5600	5.7	0.97	3392	6.2	60.6
1990	7273	6596	17.8	0.91	3487	2.8	52.9
1991	12978	11977	81.6	0.92	4366	25.2	36.5
1992	48764	58124	385.3	1.19	11007	152.1	18.9
1993	83265	111436	90.7	1.33	27515	134.1	24.7
1994	47549	82680	-24.8	1.75	33767	31.1	40.8
1995	37011	91282	10.4	2.47	37521	11.1	41.1
1996	24556	73276	-19.7	2.98	41726	11.2	56.9
1979-96	283177	467734	42.2	1.65	174885	38.7	37.4

Notes: 1. The realized value of DFI is the value of arrived DFI. 2. the realization ratio = actually arrived DFI value / pledged DFI value.
Source: State Statistical Bureau (SSB): *Statistical Yearbook of China 1994, 1995 and 1997*; and Ministry of Foreign Trade and Economic Cooperations (MOFTEC): *Almanac of China's Foreign Economic Relations and Trade, 1984-95.*

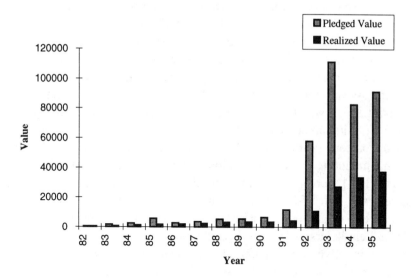

Figure 2.1 Trend of DFI in China

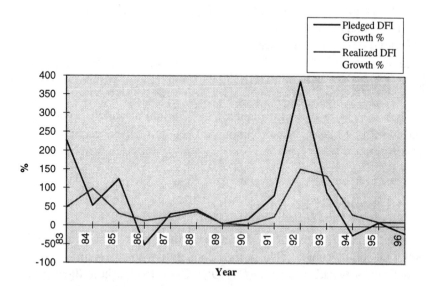

Figure 2.2   Growth Rate of DFI in China

Source: Table 2.1.

As a consequence, foreign investment has boomed unprecedentedly. As shown in Table 2.1, the pledged DFI in 1992 increased sharply by 385 percent, with a total of US$58.12 billion exceeding the cumulative total in the past 13 years (US$42,02 billion). The realised DFI also witnessed an exceptional high growth rate of 152 percent, and amounted to US$11 billion, close to the 3-year sum from 1989 to 1991. In 1993, the pledged and realised DFI continued growth by 91 percent and 134 percent respectively. The pledged DFI value amounted to US$110.9 billion, exceeding the cumulative total of the past 14 years (US$110.5 billion). The realised DFI reached to a new high, amounting to US$25.8 billion (see Table 2.1 and Figure 2.1).

Since 1994, foreign investment in China has entered a new stage of adjustment and consolidation, presenting some new features. The average capital size of foreign investment projects increases, for instance, it increased to 1.75 million dollars in 1994, 2.47 million dollars in 1995 and 2.98 million US dollars in 1996, considerably larger than in the previous years. In addition, the growth rate of DFI is back to a sustanable level from an unusually high level. From 1994 to 1996, the realised DFI increased by 14.6 percent on average per annum. Another feature is a rise in the ratio of the realized value of DFI to the pledged DFI value. For example, in 1995 and 1996, this ratio was 41 percent and 57 percent, much higher than that in 1992 (18.9 percent) and 1993 (24.7 percent).

From the recent growth trend of DFI, therefore, it can be predicted that foreign investment in China will continue to grow at a relatively high rate as long as China can maintain its political stability and futher liberalize its economy, especially make a successful progress in the economic transformation from a central planning system into a market economy. DFI would in turn be playing a more important role in Chinese economic growth and transition.

2.3 Types of DFI

The major types (forms) of DFI in China are equity joint venture (EJV), contractual joint venture (also termed cooperative joint venture, CJV) and wholly foreign-owned enterprises (WFOE). Collectively, these three forms of DFI are referred to as 'foreign-invested enterprises' (FIEs). In addition to FIEs, joint exploration of oil or coal is another form of foreign investment in China. However, this form has declined in importance since 1983.

EJV is the dominant form of foreign investment in China. An EJV involves the creation of limited liability companies with equity and management shared by foreign and Chinese sides according to their equity shares. Under the Joint Venture Law and its Amendment Provisions, a foreigner' capital contribution to a joint venture may be between 25 percent and 99 percent.

A CJV refers to an arrangement whereby the Chinese and foreign partners cooperate in joint projects and activities according to the terms and conditions stipulated in a venture agreement. These terms and conditions spell out the liabilities, rights and obligations of each partner. Unlike EJV, a CJV is not a legally independent 'entity' and does not need to form a separate 'legal person' from the partner companies. The key difference between EJV and CJV is that investment partners in the latter do not assume the risk or share profits according to their respective capital contributions. Rather, the risks and profits are predetermined by the terms and conditions laid down in the venture agreement (Grub and Lin, 1991). However, a CJV has strong similarities to an EJV. In many cases, partners in a CJV contribute capital in a various forms such as cash, building, equipment and know-how to a project that they run for a specified number of years. Before the Cooperative Joint Venture Law was published in March 1988, CJVs had been guided by equity joint venture laws (Gelatt, 1989). This resulted in a much more nebulous status in the form and content of CJV.

A WFOE is an enterprise established by a foreign company using entirely its own capital, without local Chinese capital involvement. Therefore, it assumes sole responsibility for its risks, gains and losses. The advantage of this type of enterprise is the flexibility and autonomy of investors in managing and operating the enterprise. Essentially, a WFOE is a subsidiary fully-controlled by a multinational corporation. The basic legal framework governing WFOE are the Law on Enterprises Operated Exclusively with Foreign Capital (1986) and its Enforcement Regulations (1988).

Besides the three major forms of FIEs (i.e. EJV, CJV and WFOE), the joint exploration of natural resources like oil is another form of DFI, which contains features of both the contractual joint venture and compensation trade. It involves risk sharing and distribution of output according to agreed shares, as in a CJV. It also enables China to access equipment and technical assistance from foreign companies in return for a portion of the resultant output.

Table 2.2 Foreign Investment in China by Type 1979-1996 (US$ million)

Pledged DFI	EJV	%	CJV	%	WFOE	%	J.E.	%	Total
1979-81	103	2.5	2427	59.5	296	7.3	1252	30.7	4078
1982	24	4.5	300	56.6	36	6.8	170	32.1	530
1983	188	10.9	503	29.0	40	2.3	1001	57.8	1732
1984	1067	40.2	1484	56.0	100	3.8	0	0.0	2651
1985	2030	34.2	3496	58.9	46	0.8	360	6.1	5932
1986	1375	48.5	1358	47.9	20	0.7	81	2.9	2834
1987	1950	52.6	1283	34.6	471	12.7	5	0.1	3709
1988	3134	59.2	1624	30.7	481	9.1	59	1.1	5298
1989	2659	47.5	1083	19.3	1654	29.5	204	3.6	5600
1990	2704	41.0	1254	19.0	2444	37.1	194	2.9	6596
1991	6080	50.8	2138	17.9	3667	30.6	92	0.8	11977
1992	29129	50.1	13256	22.8	15696	27.0	43	0.1	58124
1993	55174	49.5	25500	22.9	30457	27.3	305	0.3	111436
1994	40194	48.6	20301	24.6	21949	26.5	237	0.3	82680
1995	39741	43.5	17825	19.5	33658	36.9	57	- -	91282
1996	31876	43.5	14297	19.5	26810	36.6	293	0.4	73276
1979-96	217428	46.5	108129	23.1	137825	29.5	4339	0.9	467735
Realized DFI									
1979-81	65	8.8	353	47.8	1	0.1	318	43.1	738
1982	34	7.9	178	41.4	39	9.1	179	41.6	430
1983	74	11.7	227	35.7	43	6.8	292	46.0	635
1984	255	20.3	465	37.0	15	1.2	523	41.6	1258
1985	580	35.0	585	35.3	13	0.8	481	29.0	1659
1986	804	42.9	794	42.3	16	0.9	260	13.9	1875
1987	1486	64.2	620	26.8	25	1.1	183	7.9	2314
1988	1975	61.8	780	24.4	226	7.1	212	6.6	3194
1989	2037	60.0	752	22.2	371	10.9	232	6.8	3393
1990	1886	54.1	674	19.3	683	19.6	244	7.0	3487
1991	2299	52.7	764	17.5	1135	26.0	169	3.9	4366
1992	6115	55.6	2123	19.3	2520	22.9	250	2.3	11008
1993	15348	55.8	5238	19.0	6506	23.6	424	1.5	27515
1994	17925	53.1	7120	21.1	8036	23.8	678	2.0	33767
1995	19078	50.8	7536	20.1	10317	27.5	590	1.6	37521
1996	20755	49.7	8109	19.4	12606	30.2	256	0.6	41726
1979-96	90716	51.9	36317	20.8	42552	24.3	5191	3.0	174885

Notes: EJV stands for equity joint venture, CJV for contractual joint venture, WFOE for wholly foreign-owned enterprise, and J.E stand for joint exploration.

Source: MOFTEC: *Almanac of China's Foreign Economic Relations and Trade, 1984-95*; SSB: *Statistical Yearbook of China 1985-97*.

Each type of DFI has displayed a particular pattern of development in the past 18 years. In the first stage of the open-door policy (1979-85), CJV was the most important type of DFI, accounting for 55 percent of pledged DFI, while EJV accounted for only 21.6 percent of pledged DFI. Joint exploration and WFOE accounted for 18.6 percent and 3.5 percent respectively (see Table 2.2). Since 1986, as the open-door policy spread to other coastal areas, EJV has witnessed a more rapid growth than CJV. During the period 1987 to 1996, pledged DFI in EJV grew by 36.9 percent per year on average which was higher than that of CJV (26.5 percent). As a result, EJV has outweighed CJV as the principal form of DFI since 1990.

During the same period, DFI in WFOEs displayed a strong growth momentum. Pledged DFI in WFOEs increased by 56.7 percent per year on average from 1987 to 1996. (see Table 2.2). This exceptional growth can be accounted for by the improvement in legal condition for WFOE and Taiwan's investment surge. Following the 1986 Foreign Capital Enterprise Law and its Enforcement Regulations (1988), the Chinese government passed the United Tax Law for Foreign-Invested Enterprises in early 1991. Under this United Tax Law, the previous discrimination against WFOEs in tax treatment is removed and all forms of FIEs are subject to standardised tax rates and are eligible for a two-year tax holiday and a three-year tax reduction. Since 1988, Taiwanese investment in the mainland has increased rapidly. As Taiwanese investment largely took the form of WFOEs, it contributed to the growth of WFOEs.

DFI in joint oil exploration was at a standstill due to oil price drop in world markets and unfruitful exploration records. As a consequence of the differential growth rates, the relative status and importance of these types of DFI changed significantly. During the period 1987 to 1996, the share of EJV in the total pledged DFI remained around 50 percent, while the share of CJV declined to 20 percent. Since 1989, WFOE outweighed CJV and became an important mode of DFI in China. Its share in the total pledged DFI increased from 0.7 percent in 1986 to 36.6 percent in 1996.

2.4 Major Investors

Although a large number of countries have made direct investments in China, the primary sources of DFI have been highly concentrated among several large investors. These investors are Hong Kong and Taiwan, followed by the United States, Japan, Singapore, Britain, South Korea, Germany, Canada and Australia. Hong Kong (including Macau) has been the largest investor in China since 1979, and played a leading role in investment. As shown in Table 2.3, during the period 1983 to 1995 the cumulative value of actually arrived (realized) DFI made by Hong Kong was US$ 79.9 billion, accounting for 60 percent of the total DFI in China.

Taiwan ranks second only to Hong Kong although it is a recent entrant into China's market. By the end of 1995, Taiwan made direct investment of US$11.4 billion in China, accounting for 8.6 percent of total DFI. The United States is the third largest investor, accounting for 8.1 percent of the total DFI. It is followed by Japan (10.4 percent), Singapore (3.9 percent), South Korea (2.3 percent) and Britain (1.4 percent), as shown in Table 2.3.

Table 2.3 Major Investing Countries in China 1983-1995

Country /Region	Number of Projects	Pledged DFI Value	Share %	Average Size	Realized DFI Value	Share %
Hong Kong	139105	195.63	65.34	1.41	79.85	60.03
Taiwan	26626	24.13	8.10	0.91	11.44	8.60
U.S.	16234	20.53	6.86	1.26	10.72	8.06
Japan	10306	12.84	4.29	1.24	10.38	7.80
Singapore	4609	8.60	2.87	1.87	3.92	2.95
Britain	1012	5.50	1.84	5.43	1.36	1.02
S. Korea	4247	3.78	1.26	0.89	2.26	1.95
Canada	2178	2.95	0.99	1.35	0.74	0.56
Germany	888	2.59	0.87	2.92	1.26	0.95
Australia	1848	2.08	0.69	1.13	0.74	0.56
Other	19685	22.75	7.60	1.16	10.34	7.77
Total	221253	299.38	100.0	1.35	133.01	100.0

Note: The pledged DFI value are for period 1983-1994. The unit is billion dollars for pledged and realized DFI, and is million dollars for the average size of projects.

Source: MOFTEC, *Almanac of Foreign Economic Relations and Trade of China,* 1984-95 various issues; SSB, *China Statistical Yearbook,* 1994-1997, and *China's Statistics of Commerce and Foreign Economic Relations (Zhonguo Shangye Waijing Tongji) 1952-88.*

Each of the major investors has distinct characteristics. In the initial stage of the open-door policy, Hong Kong investment was dominated by assembly and processing projects. Most of the joint ventures established by Hong Kong firms were small in capital size and concentrated in labour-intensive manufacturing sectors. During the period 1983 to 1991, average capital size of Hong Kong DFI projects was US$1.12 million. As Hong Kong investments have partially shifted to some large infrastructure projects, its average capital size increased. For instance, it was US$1.51 million in 1993 and US$1.91 million in 1994.

The primary driving forces for Hong Kong investment in China are the structural transformation from labour-intensive industries to technology-intensive industries in Hong Kong and the increasing openness of China's domestic market. Because of the rapid growth of labour costs in Hong Kong since the later 1970s, labour-intensive manufacturing industries have lost their traditional advantage and competitiveness (see Bello and Rosenfeld 1990). This has spurred the transfer of labour-intensive export industries from Hong Kong to southern China, particularly Guangdong province, where geographic and cultural proximities to Hong Kong facilitate investment activities. This has been reflected in the shift of Hong Kong manufacturing industry to southern China. By 1991, 36 percent of Hong Kong's manufacturing industry had been moved across the border to the Pearl River Delta (Baldinger 1992, p. 14). It is estimated by Hong Kong Industrial Chamber that over 80 percent of Hong Kong labor-intensive manufacturing industry has been relocated to southern China by 1996 (Interview in 1997).

Taiwanese investment in China was officially prohibited by the Taiwan government before 1987. Since the promulgation by the Chinese government of the Regulations on Encouragement of Investment by Compatriots from Taiwan in 1988, Taiwan's investment in the mainland has grown dramatically. Pledged DFI by Taiwanese firms increased from US$520 million in 1988 to US$5.4 billion in 1994 (Chung and Chai 1993; MOFTEC *Almanac* 1995), and realised DFI increased from US$222 million in 1988 to US$3.1 billion in 1995 (SSB *Yearbook* 1996).

Taiwanese investment projects exhibit three main characteristics. First, they are centred in labour-intensive export manufacturing industries such as electronic and electrical appliances, plastic and rubber products, bicycles, food processing and beverage, footwear and toys, textiles, garments and small service industries. Accordingly, the average size of investment projects is smaller than the average size of all DFI projects (Table 2.3). The first choice of destination for many Taiwanese investors is the mainland of China where the labour costs are only one-third of those in Taiwan (Chung and Chai 1993; Bello, and Rosenfeld 1990; and Economist Intelligence Unit 1995). Second, due to geo-cultural proximity, Taiwan's investment is largely located in Fujian province. Third, wholly-owned enterprise is the primary mode of Taiwan's investment in China. For example, investment in wholly-owned enterprises accounted for 64.1 percent and 65.8 percent of the total pledged investment by Taiwan firms in 1989 and 1990 respectively (MOFTEC *Almanac*, 1990, 1991).

The principal reasons for the rapid growth of Taiwanese investment in China are similar to those driving Hong Kong investment, that is, the increase in labor costs and industrial transformation from labor-intensive to capital and technology-intensive production. This has forced Taiwanese and Hong Kong manufacturers to shift labour-intensive production to regions where labour costs are lower. Hong Kong and Taiwanese investments in China are largely oriented to exports. In addition, geographic proximity and similarities in regional Chinese culture promote Hong Kong investment in Guangdong and Taiwanese investment in Fujian. These issues will be discussed in detail below. Here we note that these patterns seem to support critics of the Kojima thesis (1973, 1975 and 1982) and support the argument that the structure of DFI reflects the structure of industry in the investing country, rather than persistent cultural attributes.

South Korea has become an important investor in China since diplomatic relations between the two countries were officially normalized in 1991. Pledged and realized DFI by South Korea amounted to US$3.78 billion and US$1.2 billion respectively by 1994. South Korea surpassed Britain, Germany, Canada, Thailand and Australia and became the sixth largest investor in China. If the current trend continues, South Korea's investment status in China will rise further in the future. The characteristics of South Korean investment appear similar to Hong Kong and Taiwan; Table 2.3 shows the average size of Korean projects was less than one million US dollars, and smaller than the overall average.

In recent years, investment by Southeast Asian countries has also grown rapidly. These countries include Singapore, Malaysia, Thailand and Indonesia. The investors from these countries are mainly overseas Chinese businessmen. Ancestral ties and the lure of long-term profits are major forces driving these overseas Chinese businessmen into China's investment rush. Their investments share some common features with Hong Kong and Taiwanese investments, notably the small size of their investment projects and preference for close cultural connections. However, they also have distinguishing features. First, Southeast Asian investors are motivated not so much by China's cheap labor cost and abundant land, which are already available in their own countries, but more by China's expanding consumer demand. Second, their investment flows mainly into China's primary industries and infrastructure construction. Third, these businessmen tend to locate their projects in their ancestral towns and provinces, and as their ancestral origins are scattered widely in China, these overseas Chinese investors advance into towns or provinces often ignored by other foreign investors. Fourth, overseas Chinese are reputedly cash-rich and debt-free, and therefore able to commit to long-term investment, rather than seeking the quicker profits to be had from export processing (Selwyn 1993).

In contrast, the patterns of investment by the United States and other Western countries have differed significantly from those of Hong Kong, Taiwan, and Korea. First, investments by these industrialized countries were largely in capital and technology intensive industries, and were generally of high technology content. Consequently, the average sizes of American and European investment projects are larger than those of Asian investments. During the period 1983 to 1994, the average investment size was US$1.26 million for U.S. DFI projects, US$5.43 million for British DFI projects and US$2.92 million for Germany DFI projects (seeTable 2.3).

Second, most investment projects are aimed at the Chinese domestic market. China is the largest consumption market in the world. China's economic reforms, trade liberalization and increasing openness of the domestic market to foreign investment are important reasons for U.S. and European investments. Access to China's huge consumer population is the principal motivation for many multinational corportions (MNCs) from the United States and other industrialized countries. This is to some extent different from Hong Kong and Taiwan investments, which are largely oriented to exports using cheap labour.

Third, rather than primarily locating investment projects in the SEZs or other coastal areas, many American and European companies prefer to put their projects in major industrial and commercial centres such as Beijing, Shanghai, Tianjin, Guangzhou, Wuhan and Shenyang. For example, during the period 1979 to 1989 only 47 percent of U.S. equity joint ventures were located in the coastal region, much lower than that of Hong Kong (67 percent) (US-China Business Council 1990). Finally, the duration of U.S. joint ventures is longer. According to a database of 2491 EJVs prepared by the U.S.-China Business Council (1990), the average duration of U.S. EJVs is 15 years. This is almost two years longer than the average duration of all sampled EJVs.

Finally, Japanese investors, while interested in the Chinese domestic market, have placed less emphasis than U.S. and European investors on manufacturing and more on various forms of property development. Although Japanese investment is diversifying into capital intensive industries such as electrical equipment, electronics, precision machinery and transportation equipment, the weight of labour-intensive and resource-based industries remains high. In recent years, as the Chinese government has relaxed restrictions on the industrial scope of DFI, Japanese investment in the wholesaling, retailing and warehousing industries has increased remarkably. In terms of regional distribution, Japanese investment has concentrated in northeastern China and east coast cities such as Dalian, Tianjin, Beijing, Qingdao, and Shanghai. As the costs of labor and land have risen in these cities, Japanese investment has spread gradually to nearby areas. Many Japanese investment projects have been of small and medium scale. This is different from US and European investments.

2.5   Regional Distribution of DFI

An important characteristic of DFI in China is its concentration in the coastal region. Table 2.4 compares pledged and realised DFI in the 1980s with the 1990s. In the 1990s, DFI gradually spread over the vast inland regions, resulting in a small increase in the share of DFI in the central and western regions. However, the overall pattern of the regional distribution of DFI remained virtually the same as in the 1980s. During the period 1979 to 1989, 91.9 percent of total pledged DFI in China was located in the coastal region, and despite the huge increase in the total in the 1990s, this proportion declined only slightly, to 88.8 percent. The proportion which flowed into the inland regions was very small. During the 1980s, the share of pledged DFI in the central and western regions was only 3.6 percent and 4.5 percent respectively. DFI flowed into the central region increased both in amount and share from 1990 to 1996, compared with a decline in the western region's share in DFI.

The realised DFI has the same distribution pattern; 90.7 percent of realised DFI flowed into the coastal region in the 1980s, and 88.1 percent in the 1990s, while the central and western regions collectively received a mere 9.4 percent and 11.6 percent respectively in the two periods. Although the share of the central region in total DFI displays an increasing trend over time, it is almost offset by the decline in the western region's share. This indicates that the regional spread of DFI from the coastal region to the inland regions is making only slow progress.

Within the coastal region, Guangdong Province is the largest recipient of foreign investment. It cumulatively received DFI of US$51 billion from 1983 to 1996, accounting for 30.2 percent of the total realised DFI in China. Jiangsu was second only to Guangdong in receiving foreign investment, accounting for 11.3 percent of the total realised DFI. Fujian is in the third position, accounting for 10.4 percent, followed by Shanghai (8.6 percent), Shandong (6.7 percent), Liaoning (4.4 percent), Beijing (4 percent) and Tianjin (3.3 percent).

There are two major reasons for the imbalance of the regional distribution of DFI in China. First, the open-door policy has been oriented to the coastal region since 1979. The Chinese government initiated the open-door policy by establishing four Special Economic Zones (SEZs) where special policies favourable to foreign investors were implemented. In 1984, the government opened 14 coastal cities and granted them similar policies to the SEZs. Since then there has been a significant shift in DFI from the SEZs to other open coastal areas. In 1986, the Chinese government extended the 'open region' to the 'three deltas' including the Pearl River Delta, the Minnan Delta (the south of Fujian) and the Changjiang (Yangzi River) Delta. In 1988, Hainan Island became the fifth SEZ. In the early 1990s, Pudong New Area in Shanghai became a new focus of foreign investment, with support by government preferential policies. Although the central government has encouraged inland provinces to open up to foreign investment, the regional emphasis of the opening policy is still in the coastal region. Local initiatives have reinforced the thrust of central government policies (see Crane 1990; Zweig 1995).

Second, the coastal region has advantages over the inland regions in economic conditions and investment environment. According to Dunning's international production theory, location-specific factors account for a particular pattern of locational distribution of foreign investment (Dunning 1977, 1981 and 1988). It is a common feature in most developing countries that the extent of the spatial diffusion of foreign investment is limited, and most of these investments concentrate in major economic centers or more developed regions (Leung 1990). This is because economic and social infrastructures in major economic centers are more developed than other regions.

Table 2.4 Regional Distribution of DFI in China (US$ million)

Regions	Pledged DFI 1979-89 Value	%	Pledged DFI 1990-96 Value	%	Realized DFI 1983-89 Value	%	Realized DFI 1990-96 Value	%
Costal	26558	91.9	384413	88.8	10225	90.7	138855	88.1
Guangdong	15119	52.3	123834	28.6	5141	45.6	45829	29.1
Fujian	2219	7.9	42955	9.9	757	6.7	16889	10.7
Jiangsu	748	2.6	51200	11.8	324	2.9	18807	11.9
Zhejiang	466	1.6	16332	3.8	160	1.4	5341	3.4
Shanghai	2654	9.2	41331	9.5	944	0.8	13519	8.6
Shandong	682	2.4	27173	6.3	646	5.7	10623	6.7
Hebei	345	1.2	9837	2.3	72	0.6	2490	1.6
Beijing	1847	6.4	17613	4.1	1256	11.1	5544	3.5
Tianjin	479	1.7	15069	3.5	287	2.5	5358	3.4
Liaoning	927	3.2	19488	4.5	379	3.4	6975	4.4
Guangxi	524	1.8	8323	1.9	202	1.8	3272	2.1
Hainan	548	1.9	11258	2.6	292	2.6	4208	2.7
Central	1038	3.6	33448	7.7	470	4.2	13388	8.5
Heilongjiang	198	0.9	3773	0.9	97	0.9	1749	1.1
Jilin	75	0.3	3644	0.8	16	0.2	1451	0.9
Shanxi	36	0.2	2362	0.5	18	0.2	382	0.2
Henan	265	1.1	5483	1.2	126	1.1	1795	1.2
Hubei	171	0.7	6560	1.5	78	0.7	2723	1.7
Hunan	105	0.5	4122	0.9	58	0.5	2140	1.4
Jiangxi	99	0.4	2910	0.7	37	0.3	1186	0.8
Anhui	51	0.2	3666	0.8	29	0.2	1691	1.0
Inner Mongolia	38	0.2	928	0.2	12	0.1	271	0.2
Western	1321	4.5	14328	3.3	584	5.2	4948	3.1
Shaanxi	1007	3.5	2877	0.7	344	3.0	1241	0.8
Sichuan	178	0.6	7747	1.7	124	1.1	2601	1.7
Other[1]	136	0.5	3704	0.9	116	1.0	1106	0.7
Total[2]	28892	100	432806	100	11279	100	157596	100

Source: MOFTEC, *Almanac of Foreign Economic Relations and Trade of China, 1984-95;* SSB, *China Statistical Yearbook 1984-1997;* and *China Regional Economy: A Profile of 17 Years of Reform and Opening-Up.*

Notes: 1. The other provinces in the Western region include Xinjiang, Ningxia, Gansu, Qinghai, Yunnan and Guizhou. 2. The 'Total' here refers to the total DFI registered in all regions. It does not include DFI introduced by government ministries, which was less than 2 percent of the national total in 1990-96.

In terms of the transaction cost theory (Williamson 1973, 1979 and 1981), the advanced facilities help the investors to reduce information and other relevant costs by improving the efficiency of production and marketing. In the Chinese case, the coastal region and major economic centers such as Beijing, Shanghai, Guangzhou and Tianjin are more developed in their industrial facilities education and communication systems than the inland regions. Traditionally, the coastal region was more developed than the inland regions in economic structure, industrial infrastructure, public utilities and cultural facilities. In the past 18 years, the coast-oriented opening policy and economic reforms have further improved the investment environment of the coastal region. As a result, the gap between the coast and the inland in economic conditions and investment environment has widened. Thus, the concentration of DFI in the coastal region, especially major economic centers, can be attributed to the cost advantages associated with superior physical and social infrastructures and liberalized economic conditions.

Apart from the general characteristics of regional distribution of DFI, as discussed above, the major investing countries (regions) also show different spatial patterns. As Table 2.5 shows, Hong Kong investment concentrates in the Southeast coastal region, with 41.7 percent in Guangdong alone and 10.9 percent in Fujian over the period from 1987 to 1993. Taiwanese investment concentrates in Fujian, with 19.1 percent, followed by Jiangsu (18 percent), Guangdong (13.6 percent), and Shandong (8.2 percent). American investment spreads more widely, with a focus on the major coastal provinces or cities such as Jiangsu (16 percent), Guangdong (13 percent), Shanghai (11.1 percent), Shandong (11.1 percent), Beijing (10 percent), Liaoning (6.2 percent), and Tianjin (5.6 percent).

Japanese investors also have a regional focus, in contrast with the U.S. but similar to those of Hong Kong and Taiwan. They choose Liaoning in the northeastern region as their favored location; 17 percent of Japanese investment is in Liaoning province, followed by Jiangsu (13.8 percent), Shanghai (12 percent), Guangdong (11.2 percent), and Beijing (7 percent).

The spatial patterns of these countries' investments can be explained by variations in geo-cultural links, the technological nature of investment projects, and motivations for investment. Geographic and cultural proximity are the major reasons for Hong Kong investment in Guangdong, where the local people share the same language (Cantonese) and have close ethnic links with Chinese people in Hong Kong. Similarly, Taiwan and Fujian province not only are geographically adjacent to each other, but also speak the same dialect (Minnan language). In comparison to American and European investment, Hong Kong and Taiwanese investors are taking advantage of the abundant supply of cheap labor in China to produce goods for export markets. This investment strategy requires short distances from production sites to seaports for exports. This is an additional important reason for Hong Kong and Taiwan investments' concentration in Guangdong and Fujian provinces. In parallel fashion, Korea, a former Japanese colony and used by Japanese firms in the 1970s and 1980s as a source of cheap labor, is now itself investing in northern China.

Japan occupies a middle position. Geographic proximity is an important factor accounting for Japanese investment in Liaoning, Shandong and Shanghai. Liaoning, as the industrial centre of the Northeast of China, was the focus of Japan's 'informal empire' during the period from 1931 to 1945. Not only has the spatial pattern of Japanese imperialism been repeated; the distribution of Japanese investment with its concentration in the service sector in addition to industrial investment echoes the 'developmental' approach of the South Manchurian Railway, the government agency primarily responsible for Japanese economic penetration into North China from its foundation in 1906 into the 1920s and 1930s (Myers 1989; Nakagane 1989). Hatch and Yamamura (1996, pp. 121-2) note that the Dalian industrial park is located in the city which was the southern terminus of the South Manchurian Railway. They quote a Japanese official to the effect that the park will eventually house 70 to 80 small- and medium-sized Japanese firms. But this appears more a combination of geographic proximity and subsidies for otherwise less competitive Japanese companies; there seems little support for their thesis of successful 'strategic investment' by Japan.

Unlike Asian investors, American and other Western investors do not possess geographic proximity or close cultural ties with particular regions in China. Therefore the regional distribution of their investments is not significantly affected by geo-cultural factors. Rather, their investment distribution is affected primarily by industrial and technological factors and market orientation. In contrast to the labor-intensive and export-oriented investments of Hong Kong and Taiwan, American and European investments are largely in technologically advanced industries, using capital and technology intensive production methods. This requires a suitably developed industrial base and relevant linkage industries. In the case of China, large cities, especially those in the coastal region such as Beijing, Shanghai, Guangzhou, Tianjin, and Nanjing, are relatively more advanced in industrial structure and technology, and therefore fit the requirements of American and

Table 2.5   Regional Distribution of DFI (Pledged Value) in China by Investing Country 1987-1993 (US$ Millions)

	Hong Kong		Taiwan[1]		Japan		US	
	Value	%	Value	%	Value	%	Value	%
Coastal	122637	90.3	15612	87.2	6288	91.7	10188	89.0
Guangdong	56597	41.7	2432	13.6	769	11.2	1495	13.0
Fujian	14784	10.9	3419	19.1	243	3.5	396	3.4
Jiangsu	10665	7.9	3266	18.0	943	13.8	1850	16.0
Zhejiang	4936	3.6	923	5.1	177	2.6	439	3.8
Shanghai	7006	5.2	921	5.1	816	12.0	1296	11.1
Shandong	6621	4.9	1461	8.2	609	8.9	1287	11.1
Hebei	2025	1.5	306	1.7	430	6.3	299	2.6
Beijing	5392	4.0	775	4.3	481	7.0	1110	10.0
Tianjin	1932	1.4	416	2..3	272	4.0	638	5.6
Liaoning	3721	2.7	504	2.8	1181	17.0	717	6.2
Guangxi	4222	3.1	428	2.4	86	1.3	199	1.7
Hainan	4736	3.5	761	4.2	278	4.1	462	4.0
Central	9137	6.7	1612	9.0	309	4.1	928	8.1
Heilongjiang	1099	0.8	178	1.0	25	0.4	125	1.1
Jilin	667	0.5	99	0.5	66	1.0	139	1.2
Shanxi	513	0.4	75	0.4	15	0.2	65	0.6
Henan	788	0.6	286	1.6	77	1.1	121	1.1
Hubei	2376	1.7	404	2.3	54	0.8	129	1.1
Hunan	1213	0.9	214	1.2	21	0.3	66	0.5
Jiangxi	1233	0.9	169	0.9	18	0.3	62	0.5
Anhui	907	0.7	128	0.7	17	0.3	147	1.3
Inner Mongolia	341	0.3	59	0.3	16	0.2	74	0.6
Western	3789	2.8	655	3.7	161	2.3	364	3.2
Shaanxi	1267	0.9	155	0.9	72	1.1	98	0.9
Sichuan[2]	1065	0.8	231	1.3	45	0.6	100	0.9
Other[3]	1457	1.1	269	1.5	44	0.6	166	1.4
Total[4]	135820	100	17909	100	6857	100	11480	100

Sources: SSB, *Foreign Economic Statistical Yearbook*, 1979-1991 and 1994.

Notes: 1.Taiwan investment is for the period from 1989 to 1993.   2. The data for Sichuan province is for the period from 1987 to 1992. The data for 1993 is not available in the statistical yearbooks.   3. Other provinces in the Western Region include Xinjiang, Ningxia, Gansu, Qinghai, Yunnan and Guizhou.   4. The 'Total' here refers to the total DFI registered in all regions. It does not include DFI introduced by government ministries.

European investment in technological conditions and industrial linkages. In terms of investment motivations, American investments target China's domestic market. For example over 95 per cent of the output of Beijing Jeep and of Motorola's mobile phones are sold in China's domestic market (interview 1994). Making investments in highly populated large cities is an easy way to access local consumers, and thus effectively facilitate investors' marketing strategies and promote their sales in the domestic market. This is an important aspect underlining the concentration of American investment in large coastal cities and provinces.

2.6   Sectoral Composition of DFI

2.6.1   General Characteristics of Sectoral Composition of DFI

In the last decade, DFI has been concentrated in the industrial sector, real estate and public services. As shown in Table 2.6, the cumulative pledged DFI in the industrial sector reached US$219.1 billion during the period 1983 to 1995, accounting for 56.1 percent of the total pledged DFI in China. The real estate and public service sector is second only to the industrial sector, with pledged DFI of US$112.9 billion, 28.9 percent of total pledged DFI. This is followed by the commerce and catering industry (3.7 percent), the building industry (2.8 percent), the transport, post and communication sector (1.9 percent) and the agriculture sector (1.4 percent).

Over the last 18 years, the sectoral composition of DFI has changed dramatically. These changes were contingent on Chinese economic conditions, industrial policies and the investment environment in each individual sector. In the early stage of the open-door policy, DFI was concentrated in oil exploration, hotels, tourism, and assembling and processing. For example, from 1979 to 1983 the pledged DFI in oil exploration accounted for 38 percent of the total pledged DFI (MOFTEC, 1982-85). The grandiose Ten-Year Plan for industrial investment announced by then-Chairman Hua Guofeng in 1978 included the construction of 10 new oil and gas fields to supply domestic energy needs and produce foreign exchange, a central pillar of the entire 'Four Modernisations' campaign (Cheng 1980). Since 1984, the government policy has changed, and DFI has diversified its sectoral composition. The industrial sector has become increasingly important as a recipient of DFI, whereas the share of DFI in oil exploration declined sharply due to the decline of oil prices in international markets and an unfruitful exploration experience. The DFI share of the industrial sector increased from 14.6 percent in 1983 to 47.9 percent in 1987, in contrast to a sharp decline of the DFI share in oil exploration activity, from 53.3 percent in 1983 to zero in 1987 (MOFTEC, 1982-98).

DFI in real estate, tourism and relevant services retained its share, 36.2 percent on average for period from 1983 to 1987. The primary focus of the large proportion of DFI in this sector was investment in real estate (especially hotels), tourism, and associated services which the government hoped would earn foreign exchange, which was a major problem in the initial years. Further, in this early period China could not physically accommodate manufacturing investment projects due to poor infrastructure and also the lack of a well-defined set of regulations governing entry into the domestic market. Therefore, many DFI projects in this period have been described as 'foreigners serving foreigners' (US-China business Council 1990).

A severe foreign exchange crisis in 1985-86, resulting from increased imports of consumer durables, led the government to attempt to reorient trade policy in favor of imports of machinery and other goods 'more directly relevant to meeting national development goals' (Lardy 1992, pp. 58-9). The new orientation was also reflected in policies toward DFI. After 1987 the Chinese government made great efforts to change the sectoral structure of foreign investment.

On the one hand, the government began to discourage investment in tourism and real estate in favour of high technology and export-oriented manufacturing investment by using different tax rates and approval control . On the other hand, it loosened the restrictions on foreign investors' access to the domestic market and on the requirement for foreign exchange balance in joint ventures. As a result, DFI in the industrial sector increased steadily. The share of pledged DFI in the industrial sector increased to 75.9 percent in 1988, and remained over 80 percent for the following three years (see Table 2.6). In contrast, DFI in tourism, real estate and relevant services decreased sharply, from 39.7 percent in 1987 to about 10 percent in the next three years (MOFTEC, 1988-1992).

In the recent boom of DFI from 1992 to 1993, as shown in Table 3.6, a remarkable change in the sectoral composition is a sharp increase in investment in real estate. The pledged value of DFI in real estate projects totalled US$ 18.1 billion in 1992, eleven times larger than in 1991. In 1993, the pledged DFI in real estate continued to grow by 142.1 percent. As a result, the share of DFI in real estate increased to 31.1 percent in 1992 and 39.3 percent in 1993. The big jump of DFI in real estate can be attributed mainly to the 'real estate fever' which was associated with 'the housing reform' program prevailing in major cities across the country. This program is an attempt to commercialize housing allocation through the introduction of the market mechanism. This provides an alluring profit perspective for both domestic and foreign investors since demand for housing is very strong.

Table 2.6   Sectorial Composition of Pledged DFI in China 1983-1995
(in millions of US Dollars)

Sector	1983-87	1988-91	1992-93	1994-95	1983-95
Agriculture	410	673	1870	2708	5661
	(2.3%)	(2.3%)	(1.1%)	(1.6%)	(1.4%)
Industry[1]	5842	23877	83841	105547	219106
	(33.1%)	(81.0%)	(49.4%)	(60.7%)	(56.1%)
Geological prospecting	1397	2	84	54	1536
	(7.9%)				(0.4%)
Building industry	305	501	5717	4312	10834
	(1.7%)	(1.7%)	(3.4%)	(2.5%)	(2.8%)
Transport, post communication	300	345	3033	3727	7405
	(1.7%)	(1.2%)	(1.8%)	(2.1%)	(1.9%)
Commerce & catering	805	340	6051	7349	14545
	(4.6%)	(1.2%)	(3.6%)	(4.2%)	(3.7%)
Real estate, public services	6393	3010	61851	41697	112951
	(36.2%)	(10.2%)	(36.9%)	(24.0%)	(28.7%)
Sports, hygiene eduction	145	256	1421	3425	5246
	(0.8%)	(0.9%)	(0.8%)	(2.0%)	(1.4%)
Research & technical services	8	62	650	551	1270
			(0.4%)	(0.3%)	(0.3%)
Other sectors	2015	404	5044	4592	12054
	(11.4%)	(1.4%)	(3.0%)	(2.6%)	(3.1%)
Total	17668	29470	169560	173908	390606

Notes: 1.The 'industry' sector in China includes manufacturing, mining and power and water supply. 2. The numbers in brackets are the percentage of the total.

Sources: MOFTEC, *Almanac of China's Foreign Economic Relations and Trade,* 1984-1995, and SSB, *China's Statistical Yearbook* 1990-97.

DFI in the industrial sector concinued to increase rapily although its share in total pledged DFI declined slightly, to about 50 percent in 1993. According to the definition used in Chinese official statistics, the 'industry' sector consists of manufacturing, mining, power generating and water supply. Within the industrial sector manufacturing is the core industry, with light labor-intensive industries such as textiles, clothing, toys, footwear, food and beverage, bicycles, clocks and other consumer goods being the preferred areas of foreign investment. Capital and technology intensive industries accounted for only a small share. Table 2.7 shows the composition of pledged DFI in the industry sector. It is clear that the textile industry was the single largest industry using DFI from 1987 to 1992. The 'other industries' in Table 2.7 is a heterogeneous group of industries primarily light manufacturing industries such as food, beverage, footwear, clothing, bicycles and other consumer goods. This group of industries together with the textile industry are the major investment fields.

Table 2.7 Composition of Pledged DFI in the Industry Sector
(US$ million)

Industry	1984-89	%	1990	1991	1992	1990-92	%
Total	11163	100	5569	9623	32667	47859	100
Textile	1205	10.8	629	857	2733	4219	8.8
Chemical	538	4.8	307	500	1590	2397	5.0
Machinery	740	6.6	230	638	1353	2221	4.6
Transport vehicle	45	0.4	124	145	588	857	1.8
Electrical	353	3.2	187	554	1472	2213	4.6
Electronic	884	7.9	388	767	2354	3509	7.3
Precise	45	0.4	43	83	245	371	0.8
Oil exploration	276	2.5	0	92	46	138	0.3
Oil processing	787	7.1	226	12	1158	1396	2.9
Coal mining	2.5	--	13	0	17	30	0.06
Coal processing	12	0.1	0	18	16	34	0.06
Other industries[1]	6632	59.4	3424	5957	21095	30476	63.7

Notes: 1. 'Other industries' is an extremely heterogeneous catefory. It includes clothing, toys, footwear, food, beverage, clocks, furniture, bicycles, plastic products, tobacco processing, printing, metal smelting and processing. It also includes power generating and water supply.
Sources: MOFTEC *Annual Statistical Report*, 1985-1992.

The share of capital and technology- intensive industries in total DFI was smaller. This group of industries includes the chemical, machinery, electrical and electronics, transport equipment and precision instrument industries. They collectively accounted for 20-30 percent of total DFI in the industrial sector during the period 1984 to 1992. Of this group, the electronics industry played a leading role in using DFI. It was second only to the textile industry as the largest recipient of DFI. Following the electronics industry are electrical products, machinery, chemical and transport equipment industries. About 15 percent of the total DFI in the industrial sector flowed into these four industries. The remaining industries are oil exploration and processing, coal mining and processing, metal mining, smelting and processing and power generating and water supply. DFI in these mainly resource-based industries has remained stagnant since 1984.

2.6.2 *Industrial Composition of DFI by Country of Origin*

The industrial distribution patterns of DFI vary by the country of origin. Different investing countries show distinguishing features in allocating investment in industry branches. These different patterns of investment structure are determined by the industrial advantages of the country of origin and investment goals. According to the eclectic theory of international production developed by Dunning (1977, 1981, 1988), industry-specific factors and especially firm-specific advantages in technology and production efficiency are the principal determinants of the industrial structure of investment abroad. Dunning conceives of DFI as the extension of industrial advantages from domestic markets to a foreign country.

In the case of China, DFI also demonstrates the country of origin effects on the industrial structure of investment. As shown in Table 2.8, investments from Hong Kong, Taiwan and Japan mainly target labor intensive manufacturing industries such as the food and beverage, and textiles and sewing industries, while investments made by U.S. and European firms are relatively technology-intensive, with a large unit value of projects. Investments in the textiles and sewing, food and beverage, and light manufacturing industries account for relatively large share for Hong Kong (45.8 percent), Taiwan (45.2 percent), and Japan (50.5 percent), while the shares are less for American (35.2 percent) and European investors (32.4 percent). Conversely, these Asian investors have a lower share in capital- and technology-intensive industries as compared to Western investors. The investment share in the chemicals, pharmeceuticals, electronics (a large portion of this industry is labor-intensive) and machinery industries is 45.7 percent for Hong Kong, 44.1 percent for Taiwan and 39.2 percent for Japan. These contrast with the corresponding American (52.5 percent) and European (61.8 percent) shares.

The variation of industrial composition of DFI among the major investing countries can be explained in part by the comparative advantages of investing countries and their motivation to invest in China. In general, the U.S. and European countries have advantages over their Asian competitors in high-technology industries such as chemicals, electronics, pharmaceutical and machinery manufacturing industries. In labor-intensive manufacturing industries, however, Asian investors are more competitive than their Western counterparts. Therefore, it can be argued that industrial composition of DFI varies according to the comparative advantages of investing countries.

In addition, Hong Kong and Taiwan investments in China are largely oriented to exports, using Chinese labor resources. This fits Chinese industrial advantage and competitiveness. By comparison, American and European investments principally target Chinese domestic market. This is partly explained by Chinese government policies. In processing investment applications, the MOFTEC officials use the projected export to output ratio as an important criterion for approval.

Under the Chinese industrial policy orientation, capital- and technology-intensive investments are encouraged, and one aspect of that encouragement is that they are allowed access to the domestic market. Many American and European investment projects fall into this category of investments. They invest in the industries in which China has a disadvantage and needs to develop through foreign investment. In contrast, non-technology projects are limited in their right to sell in the domestic market unless their products can substitute for imports.

Given the level of China's development and the specific facts of Chinese resources, labour-intensive and export-oriented investment projects have proved quite successful in the past 18 years. This is despite the absence of formal sanctions for non-fulfillment of export requirements. However, capital- and technology-intensive investment projects were less successful compared to labour-intensive and export-oriented investment projects. Efforts to maintain the balance of foreign exchange often mean that domestic market-oriented FIEs have difficulties unless they use locally made inputs. The lack of suitably developed linkage industries, shortage of skilled labour, limited access to the domestic market, and the difficulty of balancing foreign exchange, therefore all work to inhibit these kinds of investments.

Table 2.8 DFI in China's Manufacturing Sector by Industry and Country of Origin in 1992[1] (US$ millions)

Industries	Hong Kong[2]		Taiwan		Japan		US		Europe		Other		Total	
	Value	%	Value	%	Value	%	Value	%	Value	%	Value	%	Value	%
Food & beverage	71027.5	9.0	22600	15.7	5090	13.9	13926	12.4	3691	8.7	11806	15.9	133101	11.1
Textile and sewing	176860	22.4	23712	16.5	10223	27.8	15923	14.2	6905	16.4	12669	17.0	241182	20.1
Light manufacturing	114183	14.4	18741	13.0	3229	8.8	9627	8.6	3066	7.3	17217	23.2	166063	13.8
Chemicals, plastic	150368	19.0	22772	15.8	3549	9.7	24690	22.0	11431	27.1	5401	7.3	218211	18.2
Pharmaceutical	24683	3.1	2008	1.4	789	2.1	2994	2.7	1543	3.7	8149	11.0	40166	3.3
Electronics, Machinery	186852	23.6	38701	26.9	10057	27.4	31250	27.8	13106	31.0	12873	17.3	293742	24.5
Other industries	66960	8.5	15300	10.6	3780	10.3	13932	12.4	2475	5.9	6220	8.4	108667	9.0
Total	790934	100	143834	100	36717	100	112342	100	42217	100	74335	100	1201132	100

Notes: 1. The value of investment in this table refers to pledged value (i.e. contracted value). 2. Investment from Macao is included in Hong Kong investment.

Source: MOFTEC, 1993.

2.7 Conclusions

This chapter presents an analysis of the development process and characteristics of foreign investment in China during the past 18 years. It has found that the pattern of DFI varies according to investors' sociocultural links with the host country, firms' comparative dvantages, and the technological nature and market orientation of investment projects. In addition, the host country's policies have significant impact on the structure, trend and characteristerics of foreign investment, which are contingent on the institutional context and business conditions of the host country.

This study confirms that geographic proximity and cultural and ethnic ties account for the concentration of Hong Kong investment in Guangdong, Taiwan investment in Fujian, and Japanese investment in northeastern China. It also found that Hong Kong and Taiwan investment in the mainland of China have been largely export-oriented, using the abundant and cheap labor resources. This requires the locations of investments to be close to main seaports in order to reduce transportation costs. By comparison, American and European investments use relatively advanced technologies and mainly target Chinese domestic markets. Accordingly, their investments have been located in major cities where a large size of population and better developed industrial facilities fit their investment for both production and marketing.

It also found that the industrial composition of DFI changes from one country to another according to the investing country's comparative advantage and investment motivation. Hong Kong and Taiwan investments are largely concentrated in labor-intensive manufacturing industries, which conforms with changings in their demestic industrial structure. By contrast, American and European investments focus on capital- and technology-intensive industries targeting Chinese domestic market, while Japan lies between these two extremes.

The findings give little support to overarching, single-factor theories of foreign investment, particularly those which seek to identify differences between the roles of Japanese and 'Western' investment in Asian development, or which see Japanese investment as ominously 'strategic'. Firstly, foreign investment in China is not predominantly Western and Japanese, but Hong Kong and Taiwanese. Further, the results of the current study are broadly consistent with multi-factor approaches, such as the main propositions of Dunning's eclectic theory of international production, with location-specific, industry-specific, and firm-specific factors leading to advantages for certain types of firms in certain environments. Locational factors, especially regional economic and social environments and geographic distance, largely determine the locations of investment. The investing country's industrial structure, comparative advantage and technological level also effectively influence both the composition of investment and the mode of entry.

3 Entry Modes of Multinational Corporations into China

3.1 Introduction

As the economy booms, China has become an important focus of international investment. Many multinational corporations (MNCs) increasingly target the large domestic market of China through direct foreign investment (DFI). However, for many intending MNCs, especially those from countries with cultural distance from China, questions of how to enter Chinese market and how to choose entry modes still remain indecisive. On the part of the host country, a study of MNC's entry behavior is essential for the formulation of DFI policy and assessment of the role of MNCs in economic development.

This chapter explores the entry modes of MNCs into Chinese market from socioeconomic perspectives. It firstly presents a theoretical discussion of MNC's entry modes into the host country within the transaction cost framework, and then investigates the entry modes of MNCs in the Chinese particular institutional and business environments. It examines MNC's entry modes from various dimensions, including investors' cultural backgrounds, the technological nature of investment projects, and the economic environments of different regions. A regression analysis will also be presented to estimate the effects of these factors on foreign equity share in foreign-invested enterprises (FIEs).

3.2 Theoretical Background

Entry modes are defined as the forms of capital participation in international enterprises. They are modes in which MNCs enter the intended host country through investment. In terms of property rights, entry mode is the ownership structure of a foreign subsidiary. There are two basic entry modes: wholly-owned subsidiary and joint venture. The joint venture (JV) mode can be broken into several sub-modes based on the percentage ownership of the equity: majority JV, balanced JV, and minority JV. These two entry modes can be realized by MNCs through acquisition of an existing enterprise or setting up a new enterprise in the host country.

In the last 20 years, the transaction cost theory has been broadly employed to explain MNC's international investment activities, including their entry mode choice between wholly-owned subsidiary and a joint venture. Several authors have contributed to the entry mode literature. For instance, Williamson (1979, 1981, 1985); Casson (1982), Buckley (1985), Anderson and Gatignon (1986) Gatignon and Anderson (1988), Beamish (1988, 1989), Hennart (1988, 1991 and 1992), Gomes-Casseres (1989), Tisdell (1990a and 1990b), and Allen and Lueck (1993), have probed into MNC's entry mode by using the transaction cost approach. They argue that transaction costs are major determinants of MNC's entry modes. A MNC tends to choose an entry mode that minimizes transaction costs.

This section will discuss three primary factors affecting MNC's entry modes within the transaction cost framework. These factors include sociocultural distance between MNC's home countries and the host country, technological nature of investment projects, and the institutional and business environments and policies of the host country.

3.2.1   Sociocultural Distance

Sociocultural distance refers to the difference in social culture between the home and host countries. It is often argued within the transaction cost framework that the greater the sociocultural distance, the lower the degree of equity participation that a MNC should aim for. This can be attributed to the following two factors associated with sociocultural distance. First, sociocultural distance creates enormous information needs, hence high information costs for intending MNCs. This is because in an unfamiliar cultural environment MNCs have little knowledge of the local market and business practice. Consequently, MNCs find it difficult to transfer home technologies and management techniques to an unknown operating environment. This disadvantage may be avoided by forming joint ventures with local firms and turning management partially over to local partners, who generally outperform new foreign entrants (Root 1987, Hymer, 1976, Bell, Barkema and Verbeke, 1996).

Secondly, operating in a foreign culture at a distance increases business uncertainty and unpredictability. This is likely to undervalue foreign investment, thereby resulting in a smaller investment involvement and a smaller equity share in a joint venture. Consequently, this raises MNCs' propensity to form joint ventures with local firms. A positive relationship between the sociocultural distance and joint venture as an entry mode has been found by several authors (Goodnow and Hanz, 1972, Gatignon and Anderson, 1988, Shan, 1991, Hu and Chen, 1993).

However, not all scholars agree on this proposition. Bivens and Lovel (1966) suggest that some firms react to sociocultural distance by demanding rather than avoiding ownership so they may impose their own operating methods. Such firms do not trust local management or local partners, and prefer sufficient control to 'do it their way'.

Historically, sociocultural links between China and other countries are highly related to geographical adjacency. Countries which are adjacent geographically to China have close cultural ties[1] Based on sociocultural backgrounds and geographic adjacency, foreign investors in China can be classified into three groups. The first group are the investors from Hong Kong and Taiwan. Most of them are Chinese and share the same or very similar culture with people in the Mainland China. In contrast to true 'foreigners', these Chinese investors have advantages in language, cultural traits and ethnic links, and in access to Chinese society. These advantages allow them easily to enter into Chinese domestic market with less reliance on local firms. Therefore, investors in this group are expected to be less dependent on local firms for local management and market information. In many cases, it is not necessary for them to form joint ventures with local firms with a view to reducing transaction costs associated with an unfamiliar operating environment. In the case where these investors choose joint ventures with local firms, they tend to invest a larger share in joint ventures than other groups of investors.

Investors from other East Asian countries including Japan, Singapore, Malaysia and South Korea form the second group. These Asian neighboring countries have close cultural ties to China due to geographic proximity and historical links. In contrast to Western investors who fall into the third group, investors from these East Asian countries have some advantages in sociocultural linkages. As a result, the entry modes of these investors would be similar to that of the first group and different from the third group.

3.2.2   Research and Development Intensity

Proprietary knowledge is an important type of specialized asset. It is a core component of firm's ownership advantages and influences effectively MNC's international production and entry modes. The proprietary nature of a product, process and the amount of marketing expertise firms possess are factors found to be highly correlated with percent ownership (Hu & Chen 1993, Pan 1996, Buckley and Casson 1996). In the entry mode literature, it is argued that firms seek to exert more control as the proprietary content of the product increases (Anderson & Gatignon 1986). Research and development intensity (R & D expenditure to total sales) is usually used to measure the intellectual proprietary content of a product or process.

Stopford & Wells (1972) and Coughlan & Flaherty (1983) find a negative correlation between R & D expenditure and the proportion of subsidiaries organized as joint ventures rather than wholly-owned affiliates. A higher degree of control is more often employed for technically sophisticated products, which tend to have a higher proprietary content than unsophisticated products. This implies that firms tend to employ a completely-controlled entry vehicle in order to protect their interest in proprietary knowledge. As Anderson and Gatignon (1986), and Kim and Hwang (1992) argue, entry modes offering higher degrees of control are more efficient for highly proprietary or poorly-understood products and processes. Thus, newer technology is likely to be handled by wholly-owned subsidiaries which offer high control (Williamson 1979).

On the other hand, 'the more mature the product class, the less control firms should demand of a foreign business entity' (Anderson and Gatignon, 1986, P.13). As the diffusion of technology occurs, the products gradually lose their initial proprietary value and become mature, therefore less administrative control is needed. As Williamson (1979) points out, old technology is likely to be licensed or handled by a joint venture (lower control).

In terms of bargaining power, foreign firms with proprietary products and high technology are in a favorable bargaining position with the host country. They may force the host government and partners to allow them to have more ownership (Davidson 1982). As the product matures, the advantage erodes, creating pressure to give up control. Therefore, a foreign investing firm would have a higher propensity to form joint ventures with local firms if its technology is standardized. On the part of the host country, if local firms hold significant proprietary assets and marketing expertise, they will either have fewer incentives to form joint ventures with foreign firms, or require more control over joint ventures.

3.2.3 The Host Country's Conditions, Risks and Policies

In international operations, external uncertainty is a critical factor. External uncertainty is the volatility (unpredictability) of the firm's environment. It is typically labeled 'country risk'. This can take various forms, e.g. political instability, the lack of a well-defined legal system, economic fluctuations, price and foreign exchange controls and nationalization threat. In a highly unpredictable environment, MNCs tend to limit their equity involvement by avoiding full ownership in order to diversify the business risks. At the same time, they may want to get greater control to compensate for high risks. As Anderson & Gatignon (1986) maintain, the greater the combination of country risks, the higher the appropriate degrees of control over the subsidiaries, and the lower the MNC's inclination to establish wholly-owned subsidiaries.

In terms of the transaction cost theory, country risk and the uncertainty of business environment may increase the information need and management cost for MNCs to operate fully-owned subsidiaries in the host country. To minimize transaction costs and business risks, MNCs need to form joint ventures with local firms. For investments oriented to the host country's domestic market and investments based on the host country's resources, local partners are essential for foreign entrants since they have expertise in exploring domestic markets, managing local labor and organizing local supplies of raw materials and intermediate products.

Apart from country risk and business uncertainty, the host country's economic conditions and policies may considerably affect MNC's capital involvement. The economic growth trend and domestic market size of the host country are crucial for intending foreign investors, especially for market-seeking investors. A booming economy and rapidly expanding domestic market in the host country would strengthen its attraction for foreign investment. As is found by Gomes-Casseres (1990) using logic regression analysis, economic growth in the host country is positively correlated with the establishment of joint ventures.

In addition, the government policies of the host country can substantially influence the ownership structure of foreign subsidiaries. If the host country policies encourage joint venture by using favorable policy treatments such as tax concessions, other things being equal, MNCs may have incentives to choose joint venture as an entry mode. In some cases, the host government may require MNCs to form joint venture with local firms. This is particularly true for foreign investment based on natural resources like oil or coal, because foreign investors need the permission of the host government to gain access to these natural resources. Therefore, the policy orientation of the host government is one of major determinants for the ownership of foreign subsidiaries. In many cases, the government pressure of the host country especially developing countries plays a very important role in determining the mode of foreign entry. In the case of China, the particular political, institutional and business environments have determined some unique characteristics of foreign investing ventures.

3.3 Empirical Investigation of MNCs' Entry Modes

In the Chinese case, MNCs may take the following three modes to enter into the domestic market: equity joint venture (EJV), contractual joint venture (CJV) and wholly foreign-owned enterprise (WFOE). These three types of enterprises are collectively defined in China as foreign-invested enterprises (FIEs). The three types of FIEs are different in legal form, capital and risk involvement, and management structure. An EJV is a new limited liability company created by foreign and Chinese partners with equity and management shared in a negotiated proportion.

A CJV is an arrangement whereby Chinese and foreign partners cooperate in some joint projects or activities according to the terms and conditions stipulated in a venture contract. Although it also involves foreign capital or technology, a CJV is not a legally independent 'entity' with a separate 'legal person'. Because of the absence of a clear and independent legal form and the less favorable policy treatment by the Chinese government, CJV has declined in importance since the mid-1980s. In contrast with CJV, WFOE is becoming more important as an entry mode of MNCs into Chinese market. Therefore, the present study will focus on EJV and WFOE. Some implications derived from the analysis of EJV can be applied to CJV, due to the similarities between EJV and CJV.

In this section, an empirical investigation of MNC's entry modes in China is presented using systematic statistical data. Firstly, MNC's entry modes are explored from perspectives of country group, industry, and policy treatments and risks. A multiple regression analysis testing the effects of sociocultural distance, technology nature of products and regional business risks on the entry modes of MNCs is performed.

3.3.1 Entry Mode by Country Group

Sociocultural distance between investing countries and the host country would promote MNCs to resort to joint ventures with local firms in order to enter the host country market. It may also prevent MNCs from large capital involvement in joint ventures, resulting in a lower equity share. In the context of China, all foreign investors fall into three country groups according to their sociocultural distance from the host country. Group one is Hong Kong and Taiwan, group two is other East Asian countries and group three includes all other countries.

Based on the annual statistics prepared by the Ministry of Foreign Trade and Economic Co-operations (MOFTEC), the distribution of the three types of entry modes by country over the period of 1987-92 has been calculated and presented in Table 3.1. As can be seen from this table, there are considerable differences in the ownership structure of FIEs between the three groups. The difference between the first group and the third group is particularly significant. Due to the close cultural ties between Hong Kong, Taiwan and Mainland China, investors from the first group have a lower inclination to set up EJVs in China than other groups of investors. For instance, during the period from 1987 to 1992, the proportion of DFI in EJVs to the total DFI by Hong Kong is 47 percent on average, compared to 52 percent for Japan, 59 percent for Singapore, 68 percent for the U.S. and 84 percent for Western Europe (see Table 3.1). This indicates that the larger the sociocultural distance with the host country (China), the higher the propensity for investors to use EJV as the main entry mode. Therefore, sociocultural difference positively correlates to the frequency of equity joint venture.

However, the frequency of using CJVs seems to relate inversely to the sociocultural distance. As seen in Table 3.1, the investors of the first two groups use CJVs more frequently than the third group. This is because a CJV lacks a clear and independent legal form. It is only a cooperative agreement between the Chinese and foreign firms, thereby more likely involving business uncertainties and risks. To operate a CJV efficiently, a foreign investor must be familiar with the local market and find a partner with sufficient reliability for cooperation. Therefore, for new entrants with cultural distance, CJV is not a favored mode of entry.

Table 3.1   Entry Modes of DFI (pledged) in China by Major Investing Countries 1987-1992 (US$ million)

Year	Hong Kong						United States						Japan					
	EJV	%	CJV	%	WOFE	%	EJV	%	CJV	%	WFOE	%	EJV	%	CJV	%	WFOE	%
1987	981	50	938	47	54	3	270	79	57	17	14	4	228	76	41	14	33	11
1988	2086	58	1214	34	283	8	202	61	94	28	36	11	169	66	47	19	39	15
1989	1622	50	806	25	816	25	190	38	28	6	280	56	231	53	26	6	182	41
1990	1527	39	1029	26	1388	35	265	76	18	5	63	18	113	39	42	15	135	46
1991	3524	47	1748	23	2235	30	378	69	17	3	139	25	364	46	59	7	350	43
1992	19347	46	11469	28	10715	26	2200	70	206	7	712	23	1095	50	163	8	910	42
87-92	29087	47	17204	28	15491	25	3505	68	420	8	1244	24	2200	52	378	9	1649	39

Year	Western Europe[1]						Singapore						Taiwan[2]					
	EJV	%	CJV	%	WOFE	%	EJV	%	CJV	%	WFOE	%	EJV	%	CJV	%	WFOE	%
1987	201	90	22	10	0	0	62	88	5	7	3	5						
1988	254	92	19	7	3	1	88	64	31	23	18	13						
1989	223	92	11	4	8	4	59	53	5	5	46	42	95	22	60	14	277	64
1990	129	64	12	6	61	30	30	30	31	30	42	40	216	24	89	10	585	66
1991	671	91	5	1	36	5	77	49	15	10	63	41	555	40	169	12	664	48
1992	683	77	49	5	150	17	606	61	137	14	253	25	3076	55	507	9	1961	35
87-92	2161	84	118	5	259	10	922	59	224	14	425	27	3942	48	825	10	3437	42

Note: 1. Western Europe here includes United Kingdom, France, Italy, Netherlands, Luxembourg, Belgium and Ireland. 2. The data for Taiwan is available only for 1989-1992.

Source: MOFTEC: *Annual Statistical Report on the Utilization of Foreign Capital*, 1987-91; and *Statistical Panorama of Foreign Investment Enterprises in China 1993*.

As for WFOE, investors from the first and second groups are more likely to adopt this form to invest in China than the third group. For example, over the period from 1987 to 1992, 42 percent of Taiwan's investment and 39 percent of Japanese investment were in WFOEs, much higher than that of Western European (10 percent) and U.S. (24 percent) investments. This is because cultural similarity allows these Asian investors to have an adequate knowledge of Chinese market and business practice, and therefore facilitate their investment activities with less need for local partners. Therefore, a positive relation exists between the cultural proximity, capital involvement and the use of WFOE as MNC's entry mode.

3.3.2 Entry Mode by Industry

The technological nature and content of a product or process are highly correlated with the ownership structure of foreign subsidiaries. The higher the technological content, the higher the equity share a MNC requires in its foreign affiliates, and also the higher a MNC's propensity to set up wholly-owned subsidiaries. Therefore, for technologically sophisticated products or services, MNCs prefer full ownership or majority ownership in order to control their subsidiaries efficiently and to protect their proprietary rights. However, as technology matures and standardizes, less control and protection is needed. Therefore, a joint venture become the favored vehicle by which MNCs enter the host country market.

In the Chinese case, most foreign investment projects are labor intensive and use standardized technology. This is consistently associated with a low R & D content for many investment projects. As a result, a lower percentage of wholly foreign-owned subsidiaries in all FIEs has been found in comparison to joint ventures. As shown in Tables 3.2 and 3.3, WFOEs accounted for 26.7 percent of the total registered DFI by the end of 1993 and 26.9 percent of DFI in the manufacturing sector in 1992. Equity joint ventures accounted for 49.7 and 58.6 percent respectively for the two periods.

Since the ownership structure of FIEs is simultaneously affected by many factors, the impact of technology on the ownership structure of FIEs may become blurred. However, a positive relationship between high technology content and WFOE can still be found from the available statistical data. In high technology intensive industries, DFI largely adopt the form of WFOE. For instance, DFI in the finance and insurance industry of China over the period of 1979-93 was dominated by wholly foreign-owned subsidiaries, which accounted for 55.2 percent. The scientific services sector

Table 3.2   Entry Modes of DFI in China by Industry 1979-1993 (US$ million)

	EJVs				CJVs				WFOEs				Total	
	No.	%	Value	%	No.	%	Value	%	No.	%	Value	%	No.	Value
Agriculture¹	2125	50	994	40	1249	29	869	35	872	21	600	24	4246	2462
Industry²	83951	67	46422	55	17124	14	15051	18	23531	19	22553	27	124606	84026
Geological Exploration	41	87	19	95	2	4	0.2	2	4	9	0.7	4	47	20
Construction	2853	62	1410	43	714	16	1057	32	1036	23	817	25	4603	3284
Transport, Post & Telecommunication	1020	53	1810	64	839	44	966	34	59	3	50	2	1918	2826
Commerce, Catering & Storing	4574	52	3711	50	1515	19	2079	28	2553	29	1706	23	8742	7496
Real estate, hotel	10817	56	18679	40	3277	17	14497	32	5290	27	13254	29	19384	46431
Finance & insurance	18	58	113	45	0	0	0	0	13	42	139	55	31	252
Health care, sports & social welfare	215	60	146	27	103	29	322	60	39	11	71	13	357	539
Education, culture	1040	65	548	46	309	19	3501	29	260	16	291	25	1609	1190
Scientific services	551	62	222	51	60	7	39	9	267	30	175	40	878	435
Other	615	57	616	51	172	16	226	19	299	28	380	31	1086	1222
All Sectors	107820	64	74690	50	25464	15	35456	24	34223	20	40036	27	167507	150182

Notes: 1. This sector includes agriculture, forestry, fishing and husbandry.   2. The industry sector includes manufacturing, mining and public utilities.

Source: The State Statistical Bureau of China (SSB, 1995): *China Foreign Economic Statistical Yearbook 1994*, pp311-314.

Table 3.3 Entry Modes of DFI in China's Manufacturing Sector by Industry Branch in 1992 (US$ million)

Industries	EJVs				CJVs				WFOEs				Total	
	No.	%	Value	%	No.	%	Value	%	No.	%	Value	%	No.	Value
Food[1]	2141	75	1331	63	316	11	306	15	362	13	479	23	2819	2116
Textile, sewing	4625	69	2355	54	882	13	751	17	1190	18	1236	28	6697	4342
Light manufacture	3244	65	1661	53	704	14	560	18	1036	21	923	29	4984	3144
Chemicals[2]	3503	79	2182	66	394	9	385	12	592	13	789	24	4489	3356
Pharmaceutical	554	89	402	86	43	7	24	0.7	67	10	68	14	664	494
Machinery and electronics	4680	70	2830	53	662	10	821	15	1326	20	1725	32	6668	5376
Other Industries	3657	89	1087	70	121	3	124	8	317	8	341	22	4095	1551
Total Manufacture	30471	74	11955	59	3122	10	2971	15	4925	16	5483	27	30471	20409

Notes: 1. This industry includes food, beverage, tobacco and forage processing. 2. This industry includes chemical materials and products, rubber and plastic products.

Source: MOFTEC (1994): Statistical Panorama of Foreign Invested Enterprises in China 1993.

also shows a relatively high percentage of WFOEs (40.1 percent in value). Within the manufacturing sector, the machinery and electronics industry led other industries in establishing WFOEs. One-third of DFI in this industry are in WFOEs, which is higher than all other manufacturing industries. These facts reveal that the proprietary nature of technology and technological contents of products or services are associated with a higher frequency of wholly-owned subsidiaries.

3.3.3   Entry Mode by Region

In China, the economic conditions and policy environment vary substantially among different regions (Sun, 1995). In the Southeast coastal region, including Guangdong, Fujian and Hainan, a special policy package was granted by the central government. As economic reforms have progressed, the economy of this region has become highly liberalized and shifted from the traditional centrally-planned regime to a market economy. Consequently, economic efficiency has been improved considerably. In the past 18 years starting in 1979, this region has led the economic growth of China. Furthermore, economic liberalization and special favorable policies on foreign trade and investment significantly ameliorate the investment environment of this region, and strengthen its attraction for foreign investment. For instance, in the Special Economic Zones (SEZs) located in this region, a special preferential tax rate (15 percent) rather than a normal tax rate (33 percent) is applied to foreign investment.

In comparison with the Southeast coastal region, other regions, especially the inland regions, lack a liberalized market economic condition and preferential policies. To a great extent, the central planning system still governs the economy. The economic policies including foreign trade and investment policies in these regions are less favorable compared to the Southeast coastal region. Therefore, the business environment is rather restrictive and less attractive for foreign investment. As a result, business uncertainty and risks are higher in the these regions than in the Southeast coastal region.

The economic conditions and business risk levels in different regions are expected to influence the entry modes of MNCs significantly. This is confirmed by the entry mode distribution of MNCs by region, as shown in Table 3.4. There are significant differences in the entry mode of DFI between the Southeast coastal region and the two other regions.

Table 3.4 Entry Mode of Arrived DFI by Region 1987-1993
(US$ millions)

Regions	EJV Value	%	CJV Value	%	WFOE Value	%	Total Value
Southeast Coast	9548	39.4	7416	30.5	7281	30.0	24245
Guangdong	6690	39.3	6527	38.4	3795	22.3	17011
Fujian	2037	36.7	747	13.4	2764	50.0	5548
Hainan	821	48.7	142	8.4	723	42.9	1686
Other Coast	16325	74.3	2424	11.0	3210	14.6	21957
Liaoning	1670	67.0	154	6.2	669	26.8	2492
Beijing	2228	90.7	154	6.3	74	3.0	2455
Tianjin	521	71.1	24	3.2	188	25.7	732
Hebei	490	73.7	54	8.1	122	18.4	665
Shandong	2697	80.7	196	5.9	448	13.4	3341
Jiangsu	4012	80.0	434	8.6	570	11.4	5016
Zhejiang	1170	75.0	317	20.4	72	4.6	1559
Shanghai	2893	65.0	920	20.7	636	14.3	4449
Guangxi	646	51.8	171	13.7	430	34.5	1248
Inland Region	3997	69.7	776	13.5	920	16.0	5734
National Total	29870	57.5	10616	20.4	11411	22.0	51936

Source: State Statistical Bureau of China (SSB, 1992): *China Foreign Economic Statistics 1979-1991*, pp.378-386; and SSB (1995): *China Foreign Economic Statistical Yearbook 1994, pp.228-295.*

As shown in Table 3.4, the percentage of EJV is impressively lower in the Southeast coastal region than in the other two regions (other coastal region and inland region). During the period from 1987 to 1993, EJV accounted for 39.4 percent of total arrived DFI in the Southeast coastal region. This is significantly lower than other coastal region (74.3 percent) and the inland region (69.7 percent). EJV is the dominant form of DFI in the provinces other than those in the Southeast coastal region. For instance, in Beijing, the Chinese capital city, DFI is overwhelmingly concentrated in EJV, which accounted for 90.7 percent of the total arrived DFI. EJV also dominated in DFI in other provinces, e.g. Shandong (80.7 percent), Jiangsu (80.0 percent) and), Zhejiang (75.0 percent), Hebei (73.7 percent) Tianjin (71.1 percent), Liaoning (67 percent) and Shanghai (65 percent).

In contrast, EJV is not the dominant form of DFI in the Southeast coastal provinces, its share in the total arrived DFI is much lower than that of other regions. It was 39.4 percent in Guangdong, 36.7 percent in Fujian and 48.7 percent in Hainan.

The variation in using EJV as an entry mode between the Southeast coastal region and other regions can be explained by different political and economic environments. China's economic reforms and opening policies were initiated in the Southeast coastal region in earlier 1980s. Under the government special economic policy package, this region has become more economically liberalized and market-oriented than any other regions. Consequently, the level of business risks and unpredictability in the region is relatively lower than in other regions. As a result, foreign investing companies have less inclination to form EJV for minimizing business risks. Accordingly, WFOE and CJV accounted for a relatively large share of the DFI in the Southeast coastal region, 30.5 percent and 30 percent respectively. In Fujian and Hainan provinces WFOE accounted for 50 percent and 42.9 percent of the arrived DFI. These are significantly higher than that in the other coastal region (14.6 percent) and the inland region (16 percent).

From these statistics it can be concluded that a liberalized economic environment and lower business uncertainty and less risks in the Southeast coastal region tend to encourage foreign investors to form wholly-owned enterprises. By contrast, in other regions, economic reforms and liberalization as well as the openness of the economy are still quite limited and at the initial stage. As a result, the business uncertainty and risks are expected to be high. This prevents foreign investors from establishing fully-owned subsidiaries in these less liberalized regions. Therefore, the difference in entry modes by MNCs among the various regions can be attributed to the regional divergence in economic environment, government policy and business risks.

3.4 Regression Analysis of MNC's Entry Modes

In order to test the effects of the major influencing factors on the entry modes of MNCs in China, a multiple regression analysis is presented in this section. In this regression analysis, the foreign equity share in FIEs is used as the dependent variable. The explanatory variables are the cultural backgrounds of foreign investors, the technology nature of products and regions with different economic conditions and policy environments.

As has been discussed in previous sections, foreign investors in China can be considered in three groups according to their cultural backgrounds. Group one includes investors from Hong Kong and Taiwan. Group two are investors from other East Asian countries and group three includes investors from all other countries, primarily the U.S., Europe and Australia. The second explanatory variable is the technology nature of products. Based on the two-digit codes of the Standard Industrial Classification published by the United Nations in 1987, all the products produced by FIEs are broadly classified into two categories: high technology products and low technology products.

The third explanatory variable is the region in which DFI is located. This variable captures the location-specific factors. According to business environments and policy treatments, two regions within China can be identified. The first region is the Southeast coastal region including three provinces: Guangdong, Fujian and Hainan. The second region includes all other provinces. This classification is consistent with the findings based on Table 3.4. The first region is more economically liberalized. and the business risks associated with the traditional economic regime are lower.

3.4.1 Methodology and Data

In this regression analysis the dependent variable is the foreign equity share (percentage) in FIEs excluding contractual joint ventures. The range of change in the dependent variable is between 25 percent and 100 percent, since foreign equity share in a FIE is legally required to be 25 percent or above. Unlike a regression using continuous (quantitative) variables, this regression analysis uses three qualitative variables as independent variables (cultural backgrounds of investors, technology of products and region in which DFI is located). Therefore, the methodology applied in this study is somehow different from the case where all variables are quantitative (for a detailed discussion, see Jacob Cohen and Patricia Cohen, 1983, Chapter 5).

Firstly, we need to code qualitative variables by using dummy variables. In the present case, the cultural background of investors can be coded as CB. The first cultural group (Hong Kong and Taiwan) is represented by CB1, the second group (other East Asian countries) by CB2. The third group is the reference group. When an investor belongs to the first group, CB1 takes 1, and CB2 takes 0. If an investor is in the second group, CB2 takes 1 and CB1 takes 0. If an investor is in the third group, both CB1 and CB2 take 0. Therefore, the three cultural groups are distinguished by two dummy variables (CB1 and CB2). Similarly, we use TECH to code technology nature of products, taking 1 for high technology products (or services) and 0 for low technology products (or services). This dummy-variable coding methodology can also be applied to regions where DFI is located. The region is coded as REGION, taking 1 for the Southeast coastal region and 0 for all other regions.

Secondly, the demonstration and interpretation of regression results will be different from that of a regression using continuously quantitative variables. In a regression on qualitative variable(s), a partial regression coefficient (B_i) indicates the amount and direction of net change in the dependent variable (expressed in units of it) resulted from a change in one unit of independent variable, but it can not give the elasticity coefficient, i.e. the percentage change of the dependent variable associated with one percent change in an independent variable. Therefore, the regression coefficients can be interpreted only in two qualitative occasions (events) associated with the dichotomy of the variables.[2]

The multiple regression is specified as follows:

$$\text{Share} = \beta_0 + \beta_1 \, \text{CB1} + \beta_2 \, \text{CB2} + \beta_3 \, \text{TECH} + \beta_4 \, \text{REGION} + e$$

where β_i is the regression coefficient for each variable; and

CB1   = 1 if the investor is in the first cultural group (Hong Kong and Taiwan),
         = 0 otherwise.
CB2   = 1 if the investor is in the second cultural group (other East Asian countries),
         = 0 otherwise.
TECH   = 1 if the product is high-technology product,
         = 0 otherwise.
REGION = 1 if the FIE is located in Guangdong, Fujian and Hainan,
         = 0 otherwise.
e   = error term.

The data used in this analysis is from *the Statistical Panorama of Foreign Investment Enterprises in China* published by The Ministry of Foreign Trade and Economic Cooperation (MOFTEC) of China in 1993. This comprehensive two-volume statistical book includes detailed information of 34,500 FIEs registered in 19 economic sectors in 1992. The scope of this study includes 25,400 FIEs in the manufacturing sector. The sample size is equal to 5 percent of all FIEs in the manufacturing sector, i.e. 1270 FIEs. These FIEs are chosen by computer using random sampling technique.

3.4.2 Result and Interpretation

The regression is run by using the SPSS computer program, and the result is presented in Table 3.5. As illustrated, all the four independent variables positively and significantly affect the dependent variable--foreign equity share in FIEs. Firstly, the cultural links between investing countries (or region) and the host country (China) positively affect the foreign equity share in FIEs. As the coefficients of CB1 and CB2 indicate, the equity share held by investors from Hong Kong and Taiwan, and from other East Asian countries are 11.38 percentage points and 9.98 percentage points respectively higher than that by investors from the third group countries. This further confirms the hypothesis that cultural proximity is positively correlated with the wholly-owned foreign enterprises and the foreign equity share in FIEs. In other words, the larger the cultural distance between investors and China, the lower the foreign equity involvement in FIEs.

Secondly, high technology positively related to the foreign equity share in FIEs. If the investment project is high-technology intensive, the equity share held by the foreign investor is 9.31 percentage points higher than that of a low technology project. This supports the proposition raised in the first section, that the higher the technological content of project, the higher the ownership percentage a MNC should aim at, and also the higher the frequency of wholly-owned subsidiaries in the host country.

Thirdly, economic environment and government policy is the most important determinant for the entry mode (i.e. the ownership of FIEs) of MNCs. The effect of this factor can be measured by using the dummy variable 'region', since the economic environments and policies are considerably different between regions. The regression coefficient indicates that there is a significant difference between the Southeast coastal region and other regions in the foreign ownership of FIEs. In the Southeast coastal region, foreign investors prefer majority joint ventures or wholly-owned subsidiaries rather than minority joint ventures. On average, the foreign equity share of FIEs in this region is 18.5 points higher than that of FIEs in other regions.

Table 3. 5 **Multiple Regression Analysis using Foreign Equity Share as Dependent Variable**

Variables	Partial Coefficient	Standar d Error	t-ratios	Significanc e of t-ratio	Standardized Coefficient
CB1	11.3808	4.2407	2.684	0.0078	0.2146
CB2	9.9833	4.9426	2.020	0.0445	0.1588
TECH	9.3122	3.3558	2.775	0.0059	0.1626
REGION	18.5002	3.5028	5.282	0.0000	0.3133
Constant	43.0755	3.9352	10.943	0.0000	

F-stat. = 12.892, statistically significant at 1%. level (one tailed test). $R^2 = 0.6716$.

Based on the estimated regression coefficients, we can compute the average of foreign equity shares in FIEs for different country groups and different types of investment projects in different regions. The following Table 3.6 is a summary of foreign equity share in various occasions.

Table 3.6 **The Estimated Average Foreign Equity Share in FIEs**

Country Group	Southeast Coast		Other Region	
	Hi-tech	Low-tech	High-tech	Low-tech
Hong Kong, Taiwan	82.27	72.95	63.77	54.46
East Asia	80.87	71.56	62.37	53.06
Other Countries	70.89	61.58	52.39	43.08

3.5 Conclusions and Implications

This chapter resented a study of the entry modes of MNCs into Chinese market. It has found that the closer the cultural backgrounds of investors to that of China, the higher the equity share is held by foreign investors in FIEs. The sociocultural distance between the home countries and the host country discourages MNCs to invest in wholly-owned subsidiaries. Therefore, joint venture is the suitable entry mode for MNCs from a country at cultural distance from the host country.

This study also found that the technology intensity of investment projects and liberalized economic environment positively affect the foreign equity share in FIEs and MNC's capital involvement. These findings are consistent with the major propositions of the prevailing theories of multinational corporations, especially the eclectic theory (Dunning, 1977 and 1981). In particular, the results of the study confirm that the investment behavior and entry mode of MNCs are primarily determined by firm-specific advantages especially proprietary technologies and also location-specific factors.

Based on the findings of this study, some important implications can be drawn for both foreign investors and the host country. For foreign firms whose technology is standardized, equity joint venture is the most suitable mode to invest in China. This is particularly true for those investments oriented to Chinese domestic market or natural resource-based investment projects. Joint venture mechanism will help to minimize the external business uncertainty associated with the cultural distance and the lack of knowledge of local market, and therefore facilitate foreign firms' access to local markets. In addition, a joint venture may diversify business risks and produces net benefits from economies of scale.

On the part of the host country, joint venture as the dominant type of foreign-invested enterprises can contribute more effectively to the Chinese economy than other forms of DFI. This is because joint ventures provide an efficient mechanism for technology transfer from foreign investing firms to local Chinese firms. By forming joint ventures with foreign firms, local Chinese firms have more accesses to advanced equipment and technology, and learn new management skills. In addition, the joint venture mechanism facilitates development of industrial linkages between FIEs and local firms. Through input-output relations, FIEs effectively affect domestic sectors.

Other forms of DFI including CJV and WFOE are also available options for foreign investors to enter China's market. A CJV is a suitable form for project cooperation between foreign investors and Chinese companies over a short period, especially for some technical projects, infrastructures and venture investment. Since the financial and management commitments and distribution of profits between the two sides in the CJV are spelled out in the cooperation contracts, the business risks involved is limited and expectable.

A WFOE is a favored investment vehicle by MNCs investing in high technology industries, since this form of enterprise enables foreign investors to exercise a sufficient control over the operation of the business and to minimize management costs associated with joint ventures. However, one pre-condition for running a WFOE successfully is that investors (or their managers) have a sound knowledge of the legal system, market structure, business practice and economic and policy conditions of the host country. Without this condition, foreign investors would not efficiently operate a WFOE.

Notes

1   This is different from relations between the United States and Europe or between the United Sates and Latin America, as they either have similar cultural characters but are far in geographic distance or have cultural dissimilarity but are geographically adjacent.

2   Some authors (e.g. Gomes-Casseres, 1990; Hu and Chen, 1993) use logic regression approach to measure the impact of technology, cultural factor and policy on entry mode choice between joint venture and wholly-owned subsidiaries. In the current study, as the detailed data on foreign equity share in FIEs (the dependent variable) is available and the dependent variable is a quantitative variable (rather than a qualitative variable as treated in a logic regression) changing from 0.25 to 1.00, a multiple regression method is applied, using dummy variable to code explanatory variables.

4 Impact of DFI on Capital Formation, Economic Growth and Employment

4.1 Introduction

The Chinese economy has experienced rapid growth in the past 18 years, with an annual growth rate of 9.9 percent on the average. One of the driving forces for this remarkable growth has been direct foreign investment (DFI). By the end of August, 1997, the total of DFI in China amounted to US$204.4 billion, with 147 thousand foreign-invested enterprises (FIEs) being in operation. As DFI flows into China enormously, its role in Chinese economic development becomes increasingly important.

This chapter assesses the impact of DFI on domestic capital formation, economic growth, industrial production, and employment in China for the period 1979 to 1996. Some conclusions and policy implications will be drawn.

4.2 Effects of DFI on Domestic Investment and Economic Growth

DFI can affect the total capital formation of the host country in various ways. As a source of foreign capital inflow, DFI can augment total financial resources available for investment, thereby promoting capital formation in the host country. In addition, DFI can dynamically influence domestic investment by encouraging domestic investment, by creating new investment opportunities for local firms in leading industries or by easing economic growth bottlenecks such as shortages of infrastructure, investment capital and foreign exchange. DFI may take a physical form such as equipment, machinery, instruments and technology, which cannot be made locally, and therefore are essential for domestic capital formation in a developing country.

DFI can also stimulate domestic investment through industrial linkage effects, i.e. buying locally made inputs from, and providing intermediate inputs to, local firms. Furthermore, DFI may increase the host country's exports, which would have a positive effect on domestic savings and investment. However, DFI may have 'crowding out' effects on domestic investment if foreign firms compete with local firms both for the use of scarce physical and financial resources and also for product markets (Weisskopf, 1972; Stoneman, 1975; Atri & Jhun, 1990 and Warwick, 1991). These factors may affect one another, with the net effect depending on their relative strengths.

Worldwide, there are several studies on the effect of DFI on domestic investment and economic growth in host countries. In the case of Canada, Lubits (1971) and Noorzoy 1979) found that DFI had a complementary effect on the domestic investment. However, a very strong substitution effect was found for the U.S.A (Noorzoy, 1980). In a study of 23 developing countries, Areskoug (1976) concluded that DFI's effect on domestic investment in most developing countries was partially substituting. Using a simultaneous-equations model, Lee, Rana, and Iwasaki (1986) studied the effects of foreign capital inflows, including foreign private investment (FPI) and foreign aid, on economic growth and domestic savings in nine Asian developing countries. They found that FPI had a statistically significant positive effect on gross domestic product (GDP) growth and a positive, but not significant effect on domestic savings. In a recent study, Ruffin (1993) also confirms that foreign investment played a positive role in the economic growth of Asian and Pacific Region.

In the case studies of Taiwan, Tu (1990) and Schive (1990) investigated the contribution of DFI to economic development. They examined the effects of DFI on macroeconomic variables such as GDP, private fixed investment, private consumption, total exports and imports, and found that DFI stimulated private fixed investment and increased exports without significantly affecting private consumption and imports.

All the above mentioned studies investigated the impact of DFI in market economies. By comparison, China has some unique characteristics, which make an investigation of the impact of DFI more complex and difficult. China is the largest developing country undergoing economic transition from a centrally-planned system to a market economy. Previous studies of this issue in the context of market economies provide limited information for a study of the Chinese economy which is characterized by a large population size, unbalanced regional development, and a long history of self-sufficiency and isolation from the outside world. Therefore, conclusions and implications drawn from studies of small open economies have less relevance to China which has experienced a relatively short period of opening up to DFI since 1979. This increases the difficulty of acquiring appropriate time-series data over a reasonably long period for quantitative analysis.

To date, there have been few studies on the role of DFI in China. Kueh (1992) played a pioneering role in this type of study. Using the rising share of DFI in the total fixed capital investment, he confirmed that the contribution of DFI to the total capital formation in the open coastal region of China was increasingly important. Kueh also found that foreign-invested enterprises have become important players in the industrial development and export expansion of China's coastal region. However, Kueh's work lacks a rigorous econometric analysis of the impact of DFI on macroeconomic variables.

In a recent paper, Chen, Chang and Zhang (1995) assessed the role of DFI in China's post-1978 economic development. They discussed the effects of DFI on the growth of GDP, domestic savings, fixed capital investment and exports of China. They also briefly addressed the influence of DFI on China's transition to a market economy. They concluded that DFI has contributed to China's post-1978 economic growth by augmenting resources available for capital formation and promoting exports. Although a wide range of issues regarding the role of DFI in Chinese economic development and transition were discussed in their paper, the methodology and data used in the study make their conclusions less convincing and limit scope for further study. To overcome the shortcomings existing in the previous studies, the current study draws on well-defined data and a more suitable econometric model, enabling a systematic analysis of the macroeconomic impact of DFI in China to be derived.

4.2.1 The Model and Data

Due to the short period of time since the opening up of the Chinese economy in 1979, it is difficult to use time-series data at the national level to test the effects of DFI on other macroeconomic variables such as GDP growth and domestic investment. As an alternative, pooling time-series and cross-section data at the provincial level makes it possible to measure the effects of DFI. The regression model used in this study is a Kmenta Model (see Kmenta,

1986, for a detailed discussion), which is a time-series and cross-section model.

This model applies the generalized least square (GLS) method to pooled cross section and time series data, taking cross-sectional heteroskedasticity and time-wise autocorrelation into account. A regression analysis using this technique produces more reliable and efficient estimates than those using other models such as ordinary least square (OLS) or indirect least square (ILS) methods. Using this model and technique, a regression analysis to test the impact of DFI on economic growth and domestic capital formation in China can be presented. Economic growth can normally be expressed as the growth of GDP. The GDP and domestic investment equations are specified respectively as follows:

$$GDP = f \ (DS, DFI, FK, L)$$
$$DK \ = f \ (YPC, DFI, FK, EX)$$

where GDP stands for real GDP; DS for domestic savings; DFI for arrived direct foreign investment; FK for foreign capital other than DFI; L for labor force; DK for domestically financed fixed capital investment; YPC for national income per capita and EX for exports. All the economic variables including GDP, domestic savings (DS), domestically-financed fixed capital investment (DK), income per capita (YPC), DFI, other foreign capital inflow (FK), and exports (EX) are at constant prices (1983=100)[1].

In the first equation, GDP is explained by domestic savings (DS) and direct foreign investment (DFI), other foreign capital inflows (FK) and labor (L). Domestic savings and foreign investment (DFI + FK) are major determinants of capital formation and economic growth. Labor is another important factor of production. In the second equation, domestically-financed investment (DK) is a function of DFI, other foreign capital (FK), income per capita (YPC), and exports (EX). The coefficients of DFI and FK measure the effects of foreign capital on domestically-financed investment. The inclusion of income per capita (YPC) in the investment equation is justified on the grounds of the Keynesian 'acceleration principle'. Since income per capita is a basic measure of the average income level, changes in YPC would induce changes in investment but at a faster pace. The coefficient reflects the strength of the 'stimulation effect' of income growth on the investment. The inclusion of exports (EX) in the domestic investment equation reflects the 'income effect' of exports on domestic savings, and thereby on domestic investment. In addition, exports produce foreign exchange earnings, which are essential for many developing countries to import capital goods. Without imported capital goods, domestic capital formation would be handicapped since some advanced equipment and machinery cannot be produced domestically.

In order to measure directly the impact of the explanatory variables on the dependent variables (GDP and domestically financed investment) in terms of percentage change, the two equations can be expressed in logarithmic form:

$$\ln GDP = a_0 + a_1 \ln DS + a_2 \ln DFI + a_3 \ln FK + a_4 \ln L + u$$
$$\ln DK = b_0 + b_1 \ln DFI + b_2 \ln FK + b_3 \ln YPC + b_4 \ln EX + v$$

where a_0 and b_0 are the estimated intercepts respectively for the equations. The estimated elasticity coefficients a_1, a_2, a_3 and a_4 in the first equation measure the percentage change in GDP in response to a one percent change in domestic savings (DS), DFI, other foreign capital inflow (FK) and labor (L) respectively. Likewise, b_1, b_2, b_3 and b_4 in the second equation measure the percentage change in the domestically-financed investment in response to a one percent change in DFI, other foreign capital inflow (FK), income per capita (YPC) and exports (EX) respectively. u and v are stochastic error terms.

The **hypothesis** to be tested is that DFI positively influenced GDP growth and domestically-financed investment of China. Therefore, the expected signs of the parameters for DFI in both equations are positive. This hypothesis can be tested by examining the t test and the estimated regression coefficients.

The **data** used in this model are pooled cross-section and time-series data, including ten coastal provinces in China over a thirteen-year period from 1983 to 1995. Thus, each variable has 130 observations. The ten provinces are Fujian, Guangdong, Tianjin, Jiangsu, Zhejiang, Liaoning, Shandong, Beijing, Hebei and Shanghai. Hainan province is excluded from this study as Hainan was a part of Guangdong province prior to 1987, and its data was included in Guangdong's statistics. In the period under study, DFI in these ten provinces accounted for 90 percent of the national total. Therefore, the impact of DFI on the Chinese economy is concentrated in the coastal region. Furthermore, the ten coastal provinces have been the leading force in Chinese economic growth and export expansion since the early 1980s. In 1995, they accounted for 54.3 percent of the GDP and 86.1 percent of the total exports from China (SSB, 1996).

Most of the data used in this study are derived from the provincial statistical yearbooks of the ten provinces covering the period from 1983 to 1995. Some figures, especially the figures for 1995 are from national statistical materials including the statistical yearbooks of China 1996 and *Almanac of China's Foreign Trade and Economic Cooperations 1996* (MOFTEC, 1996). All the data are expressed in the 1983 constant price.

4.2.2 Regression Results and Explanation

The regression results for the growth equation and domestic investment equation are presented in Table 4.1. Since the variables are in logarithmic form, the estimated coefficients shown in Table 4.1 are the elasticity coefficients of the dependent variables (GDP and DK) in response to a one percent change in the corresponding independent variables.

Table 4.1 The Regression Results of DFI Effects

$$\ln GDP = a_0 + a_1 \ln DS + a_2 \ln DFI + a_3 \ln FK + a_4 \ln L + u$$

Independent Variables	Estimated Coefficients	t-ratio	Standardized Coefficients	Partial Correlation
ln DS	0.4995	14.65*	0.589	0.795
ln DFI	0.0472	4.566*	0.142	0.378
ln FK	0.0132	2.343*	0.036	0.205
ln L	0.3092	10.79*	0.356	0.695
Constant	2.6498	26.90*	0.000	0.923

$R^2 = 0.9458$; d.o.f.= 124; * the coefficients are statistically significant at 1% level (for a one tailed test); $F = 566.71 > F_{0.05} = 2.45$ (for a one tailed test).

$$\ln DK = b_0 + b_1 \ln DFI + b_2 \ln FK + b_3 \ln YPC + b_4 \ln EX + v$$

Independent Variables	Estimated Coefficients	t-ratio	Standardized Coefficients	Partial Correlation
ln DFI	0.1370	6.553*	0.366	0.506
ln FK	0.0100	1.021	0.023	0.091
ln YPC	0.3511	3.455*	0.287	0.295
ln EX	0.3940	8.584*	0.446	0.609
Constant	0.2393	0.352	0.000	0.031

$R^2 = 0.7428$; d.o.f. = 125; * the coefficients are statistically significant at 1% level for a one tailed test; $F = 93.84 > F_{0.05} = 2.45$ (for a one tailed test).

The results for the first equation indicate that increases in domestic savings (DS), direct foreign investments (DFI) and labor force (L) significantly contributed to the GDP growth of the coastal region. The coefficients of these variables are positive and all are statistically significant. Other foreign capital inflows (FK) positively but not significantly affected the economy.

The above results indicate that Domestic savings is the main determinant of economic growth. A 1 percent increase in domestic savings would give rise to a 0.50 percent increase in GDP. Domestic savings determine domestically financed capital formation. During the period from 1983 to 1995, the domestically financed fixed capital investment grew at 16.09 percent per annum. This created a 8.05 percent growth of GDP (16.09 x 0.50 = 8.05

Foreign capital plays an important role in Chinese economic growth. A 1 percent increase in DFI would lead to a 0.047 percent increase in GDP. During the period 1983-95, the average growth rate of DFI in the coastal region was 41.65 percent per year. This would result in a 2.0 percent (41.65 x 0.047) increase in GDP, which accounted for 16.2 percent of the GDP growth (12.37 percent). If contributions of DFI to domestic investment and exports are taken into account, the total effect of DFI on economic growth would be even more significant. Other forms of foreign capital (FK) also positively affected the economic growth.

Labor force also shows a positive and statistically significant impact on economic growth. A 1 percent increase in the labor force tends to give rise to a 0.31 percent increase in GDP.

With regard to domestic capital formation, the regression results shown that a 1 percent increase in income per capita would result in a 0.351 percent increase in the domestic investment. This is the most important determinant of the domestic capital formation. It is also found that DFI positively contributes to domestic investment. A 1 percent increase in DFI would lead to a 0.137 percent increase in domestic investment. As DFI grew at 41.65 percent on the annual average for the period 1983-95, it would induce a 5.71 percent (=0.137 x 41.65) increase in domestic investment. This is equal to 35.47 percent of the growth rate of domestically-financed investment (16.09 percent) over the same period. This indicates that one third of the increase in domestically-financed investment is directly associated with or stimulated by DFI. Therefore, the impact of DFI on domestic investment is positive and significant in the coastal region of China. In addition, exports shows a positive impact on domestic investment. A 1 percent increase in exports is expected to give rise to a 0.39 per cent increase in domestically-financed investment. During the period under study, the exports of the ten coastal provinces grew at 14.73 percent annually. This would give rise to a 5.74 percent (0.39 x 14.7) increase in domestically-financed investment, which represents a one third of the growth of domestically-financed investment .

These findings confirm a complementary relationship between DFI and the domestically financed investment, which is also observed in some studies of other Asian countries (e.g. Lee, *et al*, 1985; Tu, 1990). In the case of China, DFI can complement and stimulate domestically-financed investment in various ways. First, the formation of Sino-foreign joint ventures enables domestic investors to gain financial and physical resources for investment in new projects. Without foreign capital participation, many investment projects, particularly those requiring a large capital outlays and advanced technology, would be impossible to carry out. In addition, the joint venture mechanism offers an important channel to transfer foreign technology, management and marketing skills to local partners, with positive impacts on investment opportunities in new products.

In addition, DFI, especially in infrastructure projects such as airports, seaports, power stations, highways and other transportation and telecommunication facilities, improves the investment environment, thereby enhancing the domestic investment. In recent years, the proportion of DFI in these infrastructure projects has increased steadily. Consequently the bottlenecks for economic growth have been eased, and domestically-financed investment has been encouraged. Effectively, DFI not only adds financial resources to the host country, but also exerts a dynamic and positive effect on domestically-financed investment.

By contrast, the effect of other foreign capital inflows (mainly foreign loans) on domestically financed investment is positive but statistically less significant. This can be accounted for by the fact that some foreign official loans are matched to domestic capital while some foreign private loans may replace domestically financed investment. This results in an insignificant influence of foreign loans on domestically financed investment.

4.2.3   *Further Evidence for the Contribution of DFI to Capital Formation*

The contribution of foreign investment to the total capital formation of China can also be discussed by examining the changing share of foreign investment

Table 4.2 Sources of Finance for Total Fixed Capital Investment in China 1981-1995 (100 million yuan)

Year	State Budget		Domestic Loans		Foreign Investment		Self-raised Funds		Others[1]		Total	
	Value	%	Value	%	Value	%	Value	%	Value	%	Value	%
1981	269.8	28.1	122.0	12.7	36.4	3.8	532.9	55.4			961.0	100
1982	279.3	22.7	176.1	14.3	60.5	4.9	714.5	58.1			1230.4	100
1983	339.7	23.8	175.5	12.3	66.6	4.6	848.3	59.3			1430.1	100
1984	421.0	23.0	258.5	14.1	70.7	3.8	1082.7	59.1			1832.9	100
1985	407.8	16.0	510.3	20.1	91.5	3.6	1533.6	60.3			2543.2	100
1986	440.6	14.6	638.3	21.1	132.2	4.4	1488.5	49.3	320.0	10.6	3019.6	100
1987	475.5	13.1	835.9	23.0	175.4	4.8	1745.2	47.9	408.8	11.2	3640.9	100
1988	410.0	9.1	926.7	20.6	259.0	5.7	2416.9	54.4	457.9	10.3	4496.5	100
1989	341.6	8.3	716.4	17.3	274.2	6.6	2355.5	56.9	450.1	10.9	4137.7	100
1990	387.7	8.7	870.9	19.6	278.3	6.3	2329.5	52.4	583.0	13.1	4449.3	100
1991	373.0	6.8	1292.2	23.5	316.3	5.7	2878.6	52.3	648.8	11.8	5508.8	100
1992	334.2	4.3	2152.0	27.4	457.1	5.8	4024.6	51.2	887.0	11.3	7854.98	100
1993	436.9	3.5	2925.8	23.5	907.3	7.3	6218.8	49.9	1942.4	15.6	12457.9	100
1994	529.6	3.1	3997.6	23.5	1769.0	10.4	8388.2	49.2	3142.8	18.4	17042.9	100
1995	621.1	3.1	4198.7	21.0	2295.9	11.5	10647.9	53.2	2761.3	13.8	20019.6	100
1981-95	6067.6	6.7	19796.9	21.8	7190.0	7.9	47205.7	52.1	11601.8	12.8	90625.4	100

Note: 1. The figures for 'other' source are included in 'Self-raised funds' for the period of 1981-85.

Source: SSB (1987), *Zhongguo Guding Zichan Touzi Zilao (The Statistical Materials of China's Fixed-Asset Investment) 1950-1985*; and *the Statistical Yearbook of China*, 1991, 1992, 1993, 1994, and 1996.

in the total fixed capital investment. Since the opening up of the Chinese economy, foreign capital (especially DFI) has become an important source of finance for fixed capital investment. As Table 4.2 indicates, the share of foreign investment in the total fixed capital investment of China increased from 3.8 percent in 1981 to 11.5 percent in 1995. It has overtaken the government budget and become an important financial source for fixed capital investment in China.

In the coastal provinces the contribution of foreign investment to total capital formation is more important. For example, foreign investment constituted 16.2 percent of Guangdong's total fixed capital investment for the period from 1986 to 1992.[2] Its share in the provincial total fixed capital investment further increased to 22.4 percent of in 1994 (Guangdong Statistical Bureau, GSB hereafter, 1995). In Hainan Province, foreign investment accounted for 11.9 of total fixed capital investment in 1990 and reached 13.3 percent in 1992.[3]

Obviously, the increasing share of foreign capital in the total fixed capital investment indicates that the contribution of DFI to the total capital formation of China has become increasingly important. In addition to the direct contribution to total capital investment, DFI may induce and stimulate domestically-financed investment. The regression analysis indicates that DFI positively influences domestically financed investment in China. Therefore, the net contribution of DFI to the total capital formation in China would be more significant than is shown by its share in the total fixed capital investment. Since capital formation is the most important determinant of economic growth, as argued by development economists, DFI has promoted the economic growth of China through its contribution to domestic capital formation.

4.3 DFI and Industrial Production

Since the late 1980s foreign invested enterprises (FIEs) have become important players in industrial activities at both the national and provincial levels. By 1995, there were 49,559 FIEs established in the industry sector.[4] The gross output value of industry (GOVI) produced by FIEs has grown rapidly. In 1988, FIEs produced 15.6 billion *yuan* of GOVI (see Appendix Table 4.1). By 1995, this figure had increased to 913.8 billion *yuan* (see Table 4.3). As a result, the share of FIEs in the national GOVI of China rose from 0.9 percent in 1988 to 11.9 percent in 1995.

In terms of growth rate, FIEs' achievement is even more impressive. During the period from 1988 to 1995, the GOVI produced by FIEs grew at 65.8 percent per annum (calculated at 1988 constant prices), which is five times higher than the average growth rate of national GOVI (13.8 percent). This contrasts strikingly with the slow growth of the GOVI produced by state-owned enterprises. Over the same period, the GOVI produced by state-owned enterprises grew only by 5.7 percent per annum, which is less than half of the national average growth rate. As a result, the share of state-owned enterprises in the national GOVI declined remarkably from 56.8 percent in 1988 to 34.0 percent in 1995.[5] This clearly indicates that FIEs have become a dynamic leading force in the industry sector and play an important role in accelerating the industrial growth of China.

As for the regional distribution, the industrial activities of FIEs are largely concentrated in the coastal region where they contribute markedly to the regional industrial growth. As Table 4.3 shows, FIEs in the coastal region produced 820.7 billion *yuan* of GOVI in 1995, contributing 25.5 percent in value of the total industrial output. In some coastal provinces, FIEs have replaced state-owned enterprises (SOEs) as the dominant industrial producer. For example, in Guangdong and Fujian provinces, FIEs produced 50.3 percent and 53.7 percent respectively of the total industrial output in 1995. By comparison, the share of SOEs in the total industrial output declined to 19.2 percent in Guangdong and 23.7 percent in Fujian.

Another example is Tianjin where FIEs produced 40.4 percent of the total industrial output in 1995, and has overtaken SOEs as the most important player in the industry sector. In other coastal provinces, the contribution of FIEs to industrial production has also shown strong gains, highlighting DFI as an important driving force underpinning Chinese industrial growth, especially in the coastal region. Without FIEs' dynamic participation, it would be impossible for China to achieve the exceptional industrial growth of 14.9 percent per annum over the past 17 years (1979-95).[6]

The role of DFI in Chinese industrial development can also be examined by investigating the share of FIEs in the total value of output of various manufacturing industries. In 1995, 21.2 percent of the total value of output of the manufacturing sector was produced by FIEs. In the electronic industry and the industry group of clothing, footwear and other fabric products, FIEs have become the dominant producers, producing 60 percent and 51.5 percent respectively of the total output. Table 4.4 indicates similar figures in the precision instruments and office equipment industry (40 percent), plastic and rubber products (30.1 percent), transportation vehicles (24.6 percent) and the electrical products (24.3 Percent).

Table 4.3   The Contribution of Foreign-Invested Enterprises to China's Industrial Output[1] in 1995 by Region (Unit: 100 million *yuan*)

Regions	Total	SOEs[2]	%	FIEs	%	Others	%
Total	48342.0	19627.6	40.6	9137.9	18.9	19576.5	40.5
Coastal	32161.3	10088.0	31.4	8206.5	25.5	13866.8	43.1
Guangdong	5745.4	1103.6	19.2	2887.1	50.3	1754.7	30.5
Fujian	1337.2	317.4	23.7	718.7	53.7	301.1	22.5
Jiangsu	6915.5	1723.9	24.9	1023.3	14.8	4168.3	60.3
Zhejiang	3662.0	798.3	21.5	517.8	14.1	1316.1	35.9
Shanghai	3556.3	1265.6	35.6	1199.8	33.7	1090.9	30.7
Shandong	4137.8	1622.5	39.2	513.7	12.4	2001.6	48.4
Hebei	1780.1	861.0	48.4	157.6	8.9	761.5	42.8
Beijing	1270.2	674.8	53.1	332.4	26.2	263.0	20.7
Tianjin	1314.1	427.1	32.5	530.9	40.4	356.1	27.1
Liaoning	2340.3	1246.9	53.3	295.4	12.6	798.0	34.1
Hainan	102.6	47.1	45.9	29.8	29.1	33.0	32.2
Central	11131.7	6333.1	56.9	698.7	6.3	4099.9	36.8
Western	5049.0	3206.5	63.5	232.7	4.6	1609.8	31.9

Note: 1. The industrial out in this table refers to the gross output value produced by industrial enterprises at township and above levels. 2. 'SOEs' stand for state-owned enterprises, and FIEs for foreign-invested enterprises. The 'others' include collectively-owned enterprises, domestic private enterprises and share holding enterprises.

Sources: SSB: *Statistical Yearbook of China 1995*, pp. 378-381, and 1996, pp. 404-407.

In addition to the total value of output, indicators such as value added and pre-tax profit also show the significant contribution FIEs have made to the Chinese manufacturing sector. For example, FIEs produced 58.8 percent of the total value added and provided 60.3 percent of the total pre-tax profit in the electronics industry in 1995. Similarly, FIEs produced 50.5 percent of the total value added of the clothing, footwear and other fabric industry. In other industries, FIEs are also important producers. They produced 37.9 percent of the total value added of the precision instruments and office equipment industry, and 23.2 percent of value added of the food and beverage industry. These realities show that FIEs have become a substantial pillar of China's manufacturing sector and have played a major role in the industrial development of modern China.

Table 4.4   FIE's Role in the Manufacturing Industries of China in 1995[1] (Unit:100 million *yuan*)

Industries	Gross Output Value[2]			Value Added			Pre-tax Profit		
	Total	FIEs	%	Total	FIEs	%	Total	FIEs	%
Total	48696	10299	21.2	12220	2534	20.7	4000	706	17.7
Food, beverage	5196	1195	23.0	1062	246	23.2	360	96	26.6
Textile	4604	824	17.9	899	182	20.3	109	27	24.9
Clothing and footwear	2445	1260	51.5	549	277	50.5	97	40	40.7
Stationary and printing	1797	433	24.1	447	94	21.0	128	23	17.9
Chemicals	5591	803	14.4	1410	217	15.4	493	88	17.8
Plastic, rubber	1746	531	30.4	363	102	28.2	93	12	13.1
Metal products	66838	843	12.6	1739	171	9.8	557	44	7.9
Machinery	4122	491	11.9	1119	187	16.7	288	55	19.2
Transport vehicle	3303	813	24.6	805	189	23.5	265	35	13.1
Electrical	2594	631	24.3	604	139	23.1	176	42	23.6
Electronics	2531	1518	60.0	635	374	58.8	185	112	60.3
Instruments & office machine	426	169	39.6	123	45	36.9	26	12	44.7
Furniture	226	68	29.9	56	16	27.8	14	4	25.5
Others	5128	691	13.5	1784	186	10.4	908	43	4.7

Note: 1.The figures in this table refer to manufacturing enterprises with an independent accounting system. 2. The gross output value, the value added and pre-tax profit are in current price.

Source: SSB, *Statistical Yearbook of China* 1996, pp. 414-425.

DFI also supports the Chinese industry through technology transfer, management and labor training as well as industrial linkage effects. As many foreign-invested enterprises take the form of joint ventures, foreign investing firms combine with Chinese local firms in capital, management and technology. This facilitates technology transfer and training, and hence improves labor productivity and management efficiency. Similarly, FIEs' industrial linkages especially backward linkages with local Chinese enterprises are an important element of DFI. By buying locally made intermediate inputs, FIEs provide a strong growth stimulus to Chinese domestic sectors. As has been found in a previous study (Sun, 1996) and will also be discussed in detail in the next chapter, DFI has been concentrated in the industries with high backward linkage indices, such as electronics and electrical products, textiles, clothing, footwear, machinery and transformation vehicles. Therefore, the potential linkage effects of FIEs on the domestic sectors are significant.

As FIEs increasingly use locally made parts and materials in their production, the potential backward linkage effects are being realized. In this process, the campaign for localization of intermediate inputs used in FIEs which is supported by the Chinese government, should further promote the potential backward linkage effects of DFI on the domestic sectors to realize.

4.4   DFI and Employment Creation

Employment creation is an important benefit derived by the host country from DFI. For many developing countries, especially those countries with a large proportion of the population unemployed or underemployed, the creation of further employment is a primary policy goal. China is the largest developing country of a population of 1.2 billion. In addition to an increasing number of unemployment, underemployment and hidden unemployment are prevailing problems This is especially true in the vast rural areas and state-owned enterprises. These problems have been exacerbated by market-oriented economic reforms and privatization of state-owned enterprises.

Under these circumstances, the creation of new employment has become a top priority of the Chinese government. It is also one of the principal motivations underlying the utilization of foreign investment.

As foreign-invested enterprises have developed rapidly in recent years, their contribution to employment creation has become increasingly important. In 1995, there were over 10 million Chinese working in FIEs directly. In terms of sectoral composition, the newly created employment by DFI is concentrated in the industry sector (mainly manufacturing). In 1995, 8.94 million people were employed in FIEs in the industry sector (SSB, 1997), which represents 13.2 percent of the total employees in the sector. Within the industry sector, new jobs created by FIEs were centered in a few labor-intensive industries such as clothing, footwear and other fabric products, textiles, electronic & electrical products, and food and beverages. Of these industries, clothing, footwear and fabric products hold the single largest share of new employment. As shown in Table 4.5, 1.35 million people were employed by FIEs in this industry , accounting for 21 percent of all employees in the industry. The textile industry with 938,000 people employed by FIEs in 1995 is the second largest employer. This is followed by chemical industry (including chemical materials & products, oil refining, pharmaceuticals, and rubber products), with 649,000 people employed by FIEs. The electronics industry is another important industry of DFI, in which 631,000 people were employed by FIEs in 1995, accounting for 27.5 percent of the total employment in the industry.

Table 4.5   The Contribution of FIEs to the Total Employment in 1995
(unit: 1000 persons)

Industries[1]	FIEs	Total	FIEs	Provinces	FIEs	Total	FIE
All industries	8936	143672	6.2	Total	8936	143672	6.2
Food & beverage	654	7676	8.5	Beijing	244	2344	10.4
Textile	938	10140	9.2	Tianjin	254	2365	10.7
Clothing, footwear	1350	6555	20.6	Hebei	219	6511	3.4
Printing, paper, toy	425	5277	8.1	Liaoning	280	8013	3.5
Chemicals, rubber	649	9340	6.9	Shandong	600	11596	5.2
Plastic products	391	2626	14.9	Jiangsu	726	12022	6.0
Metal products	538	10164	5.3	Zhejiang	485	9288	5.2
Machinery	423	10247	4.1	Shanghai	571	4110	13.9
Transport vehicle	299	4793	6.2	Guangdong	2730	11481	23.8
Electrical product	445	3944	11.3	Fujian	809	4118	19.6
Electronics	631	2292	27.5	Guangxi	90	3087	2.9
Office machine	134	1123	11.9	Hainan	18	387	4.7
Mining, exploring	31	12479	0.3	Sichuan	120	10239	1.2
Furniture	116	951	12.2	Shaanxi	33	3163	1.1
Non-metal product	469	13848	3.4	Anhui	92	5985	1.5
Other	1443	42217	3.4	Others	1665	53081	3.1

Note: 1.The employment figures for the industries refer to the number of employees in all industrial enterprises in the industries.

Sources: SSB (1997), *The Data of the Third National Industrial Census of The People's Republic of China,* Beijing: China Statistical Publishing House.

The newly created employment by FIEs is distributed mainly in the coastal regions, especially Guangdong, Fujian, Jiangsu, Shanghai, Beijing, Shandong, Tianjin and Zhejiang. In Guangdong province alone, there were 2.73 million people employed by 30,000 FIEs in the industry sector in 1995, accounting for 23.8 percent of the total employment of the provincial industry sector. In Fujian province, FIEs also contributed considerably to employment creation. As shown in Table 4.5, 20 percent of the total employees in the industry sector were employed by FIEs in 1995. In other coastal provinces, the role of FIEs in creating employment is also important. For instance, in Shanghai and Beijing, FIEs provided 13.9 percent and 10.4 percent respectively of the total employment in 1995.

The positive impact of employment creation by DFI extends beyond direct employment growth. Through input-output linkage effects, especially backward linkage effects, FIEs create new demand for domestically made products, which delivers a strong stimulus to local Chinese firms. Consequently, the output and employment of the domestic sectors will increase by a multiple of DFI.

The creation of new employment by FIEs also has a considerable *income effect*. In general, the nominal salaries or wages paid by FIEs to Chinese employees are higher than those paid in the domestic sector. The increased income is expected to augment domestic consumption and savings. For example, the annual average wage of FIEs workers in 1996 was 8058 *yuan*, which was 43.3 percent higher than that in state-owned enterprises.[7]

Through creating new employment, DFI effectively facilitates labor flows from traditional sectors (e.g. agriculture) or state-owned enterprises to foreign-invested sectors or enterprises. As the traditional sectors and state-owned enterprises are characterized by a high rate of underemployment, labor flows from these sectors to FIEs result in a net increase in the overall labor productivity of the economy and a general improvement in the efficiency of resource allocation.

In addition to promoting inter-industry and inter-firm flows of labor, DFI also facilitates inter-regional flows of labor from rural areas to urban areas and from the inland regions to the coastal regions. This tends to increase the average productivity of labor in the rural areas or inland regions. Furthermore, the regional labor flows contribute to industrial growth in the urban areas or coastal regions. It can be expected that the new employment created by DFI and the resulting inter-regional or inter-industrial flows of labor, tend to improve the overall efficiency of allocation of resources, thereby creating a net gain for the Chinese economy.

4.5 Conclusions

This chapter has investigated the impact of DFI on capital formation, economic growth, industrial production and employment in China for the period from 1979 to 1995. It has found that DFI contributed significantly to Chinese domestic capital formation, industrial production and economic growth. DFI has also created a large number of new employment.

The regression analysis using pooled times-series and cross-section data indicates that DFI promoted the domestically-financed investment and economic growth of China. By contributing financial and physical capital and encouraging local investment, DFI positively contributed to the capital formation of China. In addition, DFI has advanced industrial production and has been an important catalyst driving Chinese industrial development. DFI has also played an important role in creating new employment in China especially in the coastal region and in labor-intensive manufacturing industries.

The Chinese experience with the use of DFI in economic development has valuable policy implications for other developing countries. First, as DFI can effectively influence the major determinants of economic development, such as domestic savings and capital formation, technology and productivity, employment, DFI can significantly affect the economic growth and development of a developing country. Therefore, DFI should be treated as an indispensable factor for economic development by the government of a developing country. The creation of a suitable legal, political and business environment in which DFI is operated is a necessary pre-condition if DFI is to boost the host economy positively. Finally, DFI policy should conform to the economic development goals and policy orientation of the host country and serves the strategic goals of development.

Notes

[1]   In the Chinese statistics, the data of DFI, other forms of foreign capital (FK) and exports (EX) are initially expressed in current US dollar. In this regression analysis, these data are deflated by US deflators published in US Department of Commerce (1997), *Survey of Current Business*, January, p. D-34.

[2]   Calculated from SSB: *Annual Statistical Report on the Fixed Capital Investment of China* 1987-92.

[3]   Calculated from *Hainan Statistical Yearbook* 1993, p.346.

[4]   According to the Chinese statistics, the industry sector includes manufacturing, mining and public utilities (electricity, gas, pipe water) production and supply.

[5]   Calculated from SSB: *Statistical Yearbook of China 1996*, p. 403.

[6]   The same source as note 5.

[7]   Calculated from SSB: *Statistical Year of China* 1996, pp. 122-126.

5 Comparative Efficiency of Foreign-Invested and Domestic Enterprises

5.1 Introduction

Since the mid-1980s, foreign-invested enterprises (FIEs) in China have increased markedly. By 1995, there were 233,564 registered FIEs, with 58,743 FIEs in the industry sector. FIEs have played an important role in Chinese industrial development, in 1995 producing 14.6 percent of China's total value of industrial output (SSB, 1997). In order to properly assess the impact of DFI on Chinese industry growth, a comparative study of the production characteristics and efficiency of FIEs and domestic firms, is necessary.

In recent years, several comparative studies on the efficiency and productive characteristics of foreign and domestic firms have been undertaken in the context of developing countries. For example, the study by Asheghian (1982) indicated that Iranian-American joint venture firms were more efficient than the local Iranian firms in labor productivity, capital productivity and total factor productivity. In the case of Brazil, Willmore (1986) found that in comparison to local Brazilian firms, foreign firms have a higher ratio of value-added to output, a greater export-output ratio, higher labor productivity, and a greater capital intensity.

Using a frontier production function, Sterner (1990) estimated the technical efficiency of firms having different types of ownership in the Mexican cement industry. He found that multinationality has no statistically significant effect on efficiency. Firms owned by multinational corporations were not found to be more efficient than other firms.

In the case of Taiwan, Schive (1982) and Tu (1990) have compared the differences in production behavior between domestic and foreign firms. They found that foreign firms tended to use more labor-intensive production methods than domestic firms. Tu (1990) examined the differences in production characteristics between foreign and domestic firms by using production function analysis. He found that the marginal productivity of capital in foreign firms was higher than that in domestic firms, however, the marginal productivity of labor in domestic firms was higher than in foreign firms.

In 1992, Terrell estimated the efficiency of Western capital and domestic capital in the Polish industry sector. Using the value of imported machinery and equipment as an indicator of Western capital, she found that the marginal product of Western capital was lower than that of domestic capital. Terrell concluded that an infusion of Western capital into an economic system that is not sufficiently market oriented is likely to be ineffective. Since most enterprises are state-owned and the financial markets are underdeveloped, a rapid importation of foreign capital could easily result in a large foreign debt with little improvement in economic efficiency.

In the context of China, there is a lack of similar studies. This can be attributed to a number of factors including the difficulty in obtaining reliable and sufficient data on foreign-invested enterprises, and also to the short period in which FIEs have operated in China. This chapter presents a comparative study of the production characteristics and efficiency of foreign-invested enterprises and domestic firms in China. It discusses the intensity of factors of production and accounts for the different use of capital and labor in both FIEs and Chinese domestic firms. It also provides a comparative analysis of capital efficiency in the two types of firm by estimating the incremental output-capital ratios. This is followed by an investigation of the average productivity of labor and capital for each type of firm and examination of the elasticity of output with respect to labor and capital inputs, and marginal productivity using a production function approach.

5.2. Capital Intensity

Capital intensity refers to the degree to which capital is used as an input relative to labor in production. It can be measured either as fixed capital value per unit of labor or the ratio of fixed capital value to the total wages paid to labor. Capital intensity is an important indicator measuring the technical level of the production process. In general, at given price levels of capital and labor, the higher the capital intensity the higher will be the technology content of the production. Therefore, at a given quality level (education and technical training) of labor, a high capital intensity is always associated with a high productivity of labor.

Conceptually, the combination of the two factors of production, or the ratio of capital to labor, tends to be determined by the factor endowments, prices of the factors and the technical nature of the production process. Given the condition of factor endowments and the price levels of factors, the intensity of a factor used in production depends on the technological nature of the production process. Generally, a high technology product would need more capital than that for a low technology product. For a given technological nature of production, the ratio of capital to labor is determined by the relative prices of these two factors. In terms of microeconomics, a factor should be used up to the point where its marginal cost (its purchasing price) is equal to its marginal revenue product (MC = MRP). When the price ratio of inputs equals their marginal rate of technical substitution, the combination of inputs is optimal. And since the marginal rate of technical substitution of labor for capital is MRP_L/MRP_K, it follows that the optimal combination of inputs is one where $MRP_L/MRP_K = P_L / P_K$. Put it differently, the firm should choose an input combination where $MRP_L / P_L = MRP_K / P_K$ (Mansfield, 1982, p.160).

There is a significant difference between FIEs and Chinese domestic firms in capital intensity. Based on the available statistical data, the two indicators of capital intensity (the value of fixed capital per employee and capital-labor ratio) are calculated for both domestic firms and FIEs at the national average level. As shown in Table 5.1, although the fixed capital per employee in domestic firms increased from 12.9 thousand *yuan* in 1987 to 42.2 thousand *yuan* in 1995, it is still lower than that in FIEs. During the same period, fixed capital per employee in FIEs increased from 26.4 thousand *yuan* to 68.4 thousand *yuan*, significantly higher than that in domestic firms. This suggests that labor is, on average, combined with more fixed capital in FIEs than in domestic firms, indicating that FIEs in general use more capital-intensive technologies in comparison to their domestic counterparts.

The capital/labor ratio (calculated as the ratio of fixed capital value to the total wages) is another important indicator for capital intensity. As shown in Table 5.1, there is a clear gap in the capital/labor ratio between FIEs and domestic firms. For FIEs, the capital/labor ratio is between 7.43 to 9.66, compared with the range from 5.86 to 7.96 for domestic firms. This implies the production of FIEs is generally characterized by a higher capital intensity than domestic firms.

Likewise, a significant difference in capital intensity between FIEs and domestic firms is also found in various industry branches. Table 5.2 displays the comparative information for both FIEs and state-owned enterprises (SOEs) in 28 industrial branches. With the exception of the petroleum processing industry, the capital intensity in FIEs in all the industries is higher than that of SOEs. On average in 1995, one employee in FIEs was combined with a fixed capital (net value) of 91.8 thousand *yuan*, which is 61.3 percent higher than that of SOEs (56.9 thousand *yuan*). This clearly suggests that FIEs generally use more capital and less labor in production and capital-intensive technology and production methods as compared to SOEs.

The higher capital intensity in FIEs can be explained by a greater technology content embodied in the production process and a relatively lower cost of capital (i.e. interest rate) in comparison to Chinese domestic firms. At a given factor price ratio, the nature of production technology is the primary determinant for the capital-labor ratio. In general, the production process of FIEs is more technically advanced than that in domestic firms. In terms of the factor price ratio, capital is cheaper in investing countries where capital is relatively abundant, whereas its price in China is higher because there is normally a shortage of capital. On the contrary, labor is an abundant resource in China, and the labor price is significantly lower in relation to capital. This is an important reason why Chinese domestic firms use relatively more labor and less capital than FIEs.

Another feature associated with capital intensity is the capital scale of enterprises. In general, FIEs have a relatively larger scale of capital than their domestic counterparts. As displayed in Table 5.1, each FIE had 11.9 million *yuan* fixed capital on average in 1995, which exceeded the average capital scale of domestic firms (7.11 million *yuan*). However, if we compare FIEs with state-owned enterprises in the 28 industry branches, it is found that the average capital scale of FIEs was 16.2 million *yuan* in 1995, smaller than that of state-owned enterprises (28.9 million *yuan*).

The reason for the smaller average capital size of domestic firms as a whole is the fact that 49 percent of domestic firms are township and village enterprises (TVEs), which generally have a smaller capital scale. In 1995, the average scale of fixed capital of TVEs was less than one (0.98) million *yuan* (SSB, *The Third National Industrial Census*). This results in domestic enterprises as a whole being of a smaller average capital scale compared to FIEs. The smaller capital size and the lower capital-labor ratio lead to a lower labor productivity in domestic enterprises compared to FIEs.

Table 5.1   Comparison of Efficiency Indicators between FIEs and Domestic Firms

Whole Country:	1987		1988		1989		1990		1991		1992		1995	
	DEs[1]	FIEs	DEs	FIEs	DEs	FIEs	DEs	FIEs	DEs	FIEs	DEs	FIEs	DEs	FIEs
Capital-labor ratio[2]	7.28	8.38	6.91	7.70	7.14	8.64	7.98	9.66	7.96	8.54	7.05	7.43	5.86	8.22
Fixed capital/employees[3]	12.9	26.4	1.41	2.64	1.65	3.20	19.1	43.0	21.4	54.8	24.7	45.4	42.2	68.4
IOCR[4]	1.66	2.04	1.93	1.81	1.20	1.22	0.47	1.25	1.09	1.74	0.56	0.84	0.96	1.19
Average capital productivity[5]	1.79	2.58	1.96	2.68	1.99	2.34	1.84	2.02	1.81	2.49	1.92	2.73	1.62	2.22
Average labor productivity[6]	15.9	58.6	19.3	58.4	22.9	56.2	24.6	81.8	27.3	93.6	33.4	105.1	40.4	113.0
Export/output ratio									0.10	0.44	0.09	0.38	0.11	0.39
Value-added-output ratio	0.34	0.27	0.29	0.24	0.28	0.25	0.27	0.26	0.27	0.23	0.28	0.24	0.29	0.24
Profit ratio[7] (%)	9.23	9.31	8.50	9.40	6.30	7.44	3.27	6.41	2.98	5.63	3.27	6.39	2.90	4.00
Average size of fixed capital[8]	2.19	4.67	2.53	5.74	2.97	6.93	3.45	8.91	4.10	8.73	4.78	9.77	7.11	11.9

Note: 1. 'DEs' and FIEs refer to domestic enterprises and foreign-invested enterprises respectively. 2. The capital-labor ratio = the net value of fixed capital / the total wages. 3. Fixed capital per employee = the original value of fixed capital / number of employees, the unit is 1,000 *yuan*. 4. IOCR (incremental output-capital ratio) = change in output / change in total capital (i.e. investment). 5. The average capital productivity = total output value / the total value of fixed capital used in that year, i.e. output per unit of fixed capital (the unit for both output and capital are *yuan*). 6. Average labor productivity = total output value / the number of total employees, the unit is 1,000 *yuan*. 7. profit ratio = net profit / sale revenue. 8. The unit for average size of fixed capital is million *yuan*.

Source: State Statistical Bureau (SSB): *The Annual Statistical Report of China's Industry*, 1987–92; and *The Data of The Third National Industrial Census of The People's Republic of China in 1995*, published by China Statistical Publishing House in 1997.

Table 5.2   Capital Intensity and Average Productivity of FIEs and State-Owned Domestic Firms by Industry in 1995

Industry	SOEs				FIEs			
	K/L[1]	AP_K[2]	AP_L[3]	Scale[4]	K/L	AP_K	AP_L	Scale
Food processing	95.0	2.35	95.0	7.6	104.8	3.68	331.3	10.5
Food manufacturing	32.1	1.58	43.4	5.3	93.8	1.73	141.8	10.4
Beverage manufacturing	49.1	1.59	64.0	13.1	160.2	1.57	210.0	17.2
Tobacco processing	153.4	2.95	316.5	15.6	460.0	2.75	1081.0	60.4
Textile	27.6	1.72	40.2	28.9	79.0	1.95	124.8	12.4
Garment & other fabric	21.1	2.23	40.7	4.8	24.6	4.11	82.3	3.7
Leather products	23.4	1.69	34.5	6.5	23.2	4.19	89.5	5.4
Timber processing	33.5	0.95	26.5	7.1	58.0	1.82	89.5	5.9
Furniture	24.8	1.47	33.3	2.5	49.0	2.28	92.6	4.8
Paper making	45.7	1.36	49.0	18.6	104.6	1.66	149.7	11.2
Printing	31.7	1.13	30.5	4.9	147.6	1.11	135.2	9.4
Stationery	23.2	2.06	38.2	4.6	27.5	3.63	82.1	5.3
Petroleum processing	155.0	2.10	128.0	24.2	63.1	2.20	111.7	6.1
Chemicals	61.0	1.06	51.7	33.3	127.3	2.32	245.4	10.0
Pharmaceuticals	44.4	2.02	67.0	15.5	97.3	2.48	200.0	10.6
Chemical fiber	93.2	1.24	95.2	11.1	228.4	1.07	212.6	33.0
Rubber products	40.0	2.08	60.6	25.5	56.6	2.67	126.3	14.8
Plastic products	41.9	1.50	52.9	6.9	98.4	1.83	147.5	8.3
Non-metal mineral	42.2	0.96	33.3	16.3	128.9	0.91	100.2	17.8
Ferrous metal smelting	108.3	0.69	53.4	27.6	150.3	2.27	267.0	34.1
Non-ferrous metal smelt	78.2	1.40	92.0	8.9	136.1	2.54	289.2	17.8
Metal products	28.8	1.58	37.4	7.0	105.3	2.11	187.2	10.4
General machinery	33.5	1.42	39.3	18.6	85.9	2.45	169.4	11.7
Special machinery	31.5	1.52	40.1	16.0	65.7	2.65	141.1	55.6
Transport equipment	43.0	1.89	64.8	27.7	126.0	3.54	367.1	19.8
Electrical machinery	35.4	1.79	52.5	14.9	83.2	2.72	185.5	12.7
Electronic equipment	43.3	2.09	72.3	23.6	93.1	4.05	292.0	16.7
Meter, office machinery	28.7	1.14	27.6	12.7	54.5	3.74	160.7	5.7
Overall average	56.9	1.21	58.0	28.9	91.8	2.15	162.2	12.2

Notes: 1. K/L = the total value of fixed capital / the number of employees, the unit = 1,000 *yuan*. 2. AP_K = output value / the value of fixed capital used in production, excluding those for non-production purpose. 3. AP_L = output value / number of employees. 4. The 'scale' refers to fixed capital scale each firm on average, the unit is one million *yuan*.

Source: SSB: *The Data of The Third National Industrial Census of The People's Republic of China in 1995*, published by China Statistical Publishing House in 1997.

5.3 Average Productivity of Factors

The average productivity is an important measure of the productive efficiency of labor or capital. Average labor productivity is output produced per employee on average, whereas average capital productivity is output produced by one unit of capital used in a firm. Based on the SSB data presented in the *Annual Statistical Reports on Industry* from 1987 to 1993 and *the Third National Industrial Census*, the average productivity of both labor and capital for FIEs and domestic firms are calculated and presented in Tables 5.1 and 5.2.

It is clear that the average productivity of both labor and capital in the FIEs is significantly higher than that of domestic firms. As can be seen in Table 5.1, the average productivity of labor in FIEs is two or three times that of domestic firms. In 1992 the national average output value produced per employee in FIEs was 105.1 thousand *yuan*, which was 3.14 times that of domestic firms. Similarly, a considerable difference between FIEs and domestic firms can be seen in the average productivity of capital. For instance, the average capital productivity in FIEs is 2.22 *yuan* in 1995, which was 37 percent higher than that in domestic firms (1.62 *yuan,*).

The difference in factor productivity is more significant when FIEs are compared to SOEs. As presented in Table 5.2, the labor productivity of FIEs in the industry sector[1] is apparently higher than that of SOEs. One employee produced an output value of 162.2 thousand *yuan* on average in FIEs in 1995, which is 2.78 times that of SOEs (58.0 thousand *yuan*). The average capital productivity also shows a similar pattern between FIEs and SOEs. One *yuan* fixed capital (net value) contributed to 2.2 *yuan* output value in FIEs, whereas it was associated with 1.2 *yuan* output value in SOEs. In terms of the industrial variation of average capital productivity, it was only in four sub-industries (tobacco processing, printing, chemical fiber and non-metal mineral) that SOEs outperformed FIEs. In the other 24 sub-industries, the average capital productivity in FIEs was higher than in SOEs. For labor productivity, with the exception of the petroleum processing industry, FIEs achieved a higher labor productivity in all industries compared with SOEs.

The higher average productivity of labor in FIEs can be explained by a higher capital-labor ratio (higher capital intensity), higher technical and management efficiencies. In addition, better trained workers and efficient management also contributed to the higher labor productivity of FIEs. Similarly, a higher average capital productivity in FIEs can be attributed to a greater technology content embodied in production, well-organized production processes and efficient management. It is also closely associated with labor productivity since at a given capital-labor ratio, skilled labor can produce more output. This tends to lead to greater productivity of capital.

5.4   Incremental Output-Capital Ratio

Incremental output-capital ratio (IOCR) is the ratio of a change in output to a change in capital stock (i.e. investment). It can be expressed as: IOCR = $(Y_t - Y_{t-1})$ / I, where Y is output and I is investment. Thus an IOCR indicates the multiplier effect of investment on output. The IOCR is calculated for both domestic firms and FIEs based on the information on output and investment prepared by the SSB and is presented in Table 5.1.

As depicted in Table 5.1, the IOCR in FIEs is generally higher than that in domestic firms. During the period 1987-95, the IOCR for domestic firms was 1.12 on average while the IOCR for FIEs was 1.44. This indicates that a one dollar increase in investment results in a 1.12 dollar increase in the output of domestic firms, whereas a one dollar increase in investment creates a 1.44 dollar increase in the output of FIEs. This points to the investment efficiency in FIEs being remarkably higher than that of domestic firms. A larger IOCR represents a higher productivity of capital in FIEs. It also reveals that the production process of FIEs involves more advanced technologies than that of domestic firms.

5.5   The Production Function Estimation

In order to measure the degrees of return to scale, the elasticity of output with respect to labor and capital and marginal productivity of each factor, production functions for both state-owned enterprises and foreign-invested enterprises are estimated. Using the result of production function analysis, we will shed light on the productive efficiency in the two types of firms.

5.5.1 The Models

There are three popular production functions which are widely used by researchers in production analysis. They are the Cobb-Douglas production function, the constant elasticity substitution (CES) production function, and the translog production function. The first two are homogeneous production functions and the third is a non-homogeneous production function. These three production functions will be estimated for SOEs and FIEs in this section. Based on the estimation results from all the models, the appropriate production function form will be identified. The specifications of these models are discussed briefly below.

The Cobb-Douglas production function is specified as:

$$Q = AL^{\alpha}K^{\beta}exp^{\mu} \qquad\qquad (A > 0, \alpha > 0, \beta > 0) \qquad\qquad (1)$$

where Q, L, K, μ: and t denotes output, labor input, capital input, the error term and time, respectively. A(t) is a parameter denoting the level of technology at time t. Parameters α and β are the output elasticity with regard to labor and capital respectively.

Expressing equation (1) in logarithmic form, we obtain:

$$\ln Q = c + \alpha \ln L + \beta \ln K + \mu; \qquad\qquad (2)$$

where c (= ln A) is a constant.

In this model, it is assumed that the elasticity of substitution between capital and labor is constant and is equal to unity. This implies that a one percent change in the relative prices of labor and capital will give rise to a one percent change in the relative proportion of labor and capital used in production. The Cobb-Douglas production function is homogeneous of the degree $\upsilon = \alpha + \beta$. When $\upsilon = 1$, this production function exhibits constant returns to scale, when $\upsilon > 1$, it has increasing returns to scale, and when $\upsilon < 1$, it reveals decreasing returns to scale.

Compared to the Cobb-Douglas production function, *the CES production function* is less restrictive and more general. It assumes that the elasticity of substitution between labor and capital is constant, but not necessarily unity. The original form of the CES production function is specified (see, Uzawa, 1962, Walters, 1963 and Zarembaka, 1970) as:

$$Q = A(t) [\delta L^{-p} + (1- \delta)K^{-p}]^{-(v/p)} \qquad\qquad (3)$$

where
  $A(t) > 0, 1 \geq \delta \geq 0, v > 0$, and $\rho \geq -1$.
As before, Q, L. K, and t represent output, labor, capital and time, respectively. A(t) is the level of technology at time t, δ indicates the degree to

which technology is labor intensive; v and ρ are the homogeneity and substitution parameters (Bairam, 1994). In logarithmic form, equation (3) can be transformed as:

$$\ln Q = \ln A(t) + (v/\rho)\ln[\delta L^{-\rho} + (1-\delta)K^{-\rho}] \tag{4}$$

Equation (4) indicates that a simple ordinary least square (OLS) procedure cannot be applied since it is non-linear in parameters. Alternatively, we can use its approximation which is linear with respect to ρ. The approximation of the CES model was developed by Kmenta (1967 and 1971), Maddala and Kadane (1967), who used OLS to estimate the parameters of the CES production function by replacing it with the following function:

$$\ln Q = \ln A(t) + \mu_1 \ln L + \mu_2 \ln K + \mu_3 (\ln L - \ln K)^2 + \mu_4 (\ln L - \ln K)^3 \tag{5}$$

where $\mu_1 = \delta v$; $\mu_2 = (1-\delta)v$, $\mu_3 = 0.5 [\delta(1-\delta)\rho v]$; and
$\mu_4 = \{(1/6)\rho^2\delta(1-\delta)v[1-2(1-\delta)]\}$

The right-hand side of equation (5) can be separated into two parts: one corresponding to the Cobb-Douglas production function and another representing a 'correction' due to the departure of ρ from zero. If $\rho = 0$, the two interactive items $(\ln L - \ln K)^2$ and $(\ln L - \ln K)^3$ will disappear, and the model is reduced to the Cobb-Douglas. Consequently, using this specification the Cobb-Douglas hypothesis (i.e. constant and unity elasticity of substitution between labor and capital) can be tested. If estimates μ_3 and μ_4 are not significantly different from zero, the CES production function can be rejected in favor of the Cobb-Douglas production function.

The third popular production function is *the translog production function*, which is in more general form. For the two factor case the translog production function is specified as:

$$\ln Q = A(t) + \alpha_L \ln L + \alpha_K \ln K + 0.5\alpha_{LL} (\ln L)^2 + 0.5\alpha_{Kk} (\ln K)^2 + \alpha_{Lk} (\ln L \ln K) \tag{6}$$

It is clear that the translog function is not homogeneous and, hence, the scale elasticity of labor and capital vary over time. If capital's elasticity grows relative to that of labor, the increasing weight given to capital which is the faster growing input, would cause the index of factor inputs to rise more rapidly and the index of multi-factor productivity to rise more slowly than that with constant elasticity which is specified in the Cobb-Douglas production function. Alternatively, a growing labor elasticity would cause the productivity index to rise more rapidly than with constant elasticity. (Bairam, 1994)

When parameters $\alpha_{LL} = \alpha_{Kk} = 0$, the translog production function is reduced to the Bairam-Vinod production function (Vinod, 1972 Bairam, 1994), as shown bellow:

$$\ln Q = A(t) + \alpha_L \ln L + \alpha_K \ln K + \alpha_{Lk} (\ln L \ln K) \qquad (7)$$

In order to test for the constant returns to scale assumption, as well as the Cobb-Douglas and the CES assumption, equations (6) and (7) will be estimated. If parameters for the interactive items, α_{LL}, α_{Kk} and α_{Lk}, are not significantly different from zero, the translog production function should be rejected in favor of the Cobb-Douglas production function.

5.5.2 Data

The data used in this study are cross section data at the industry level, including 189 branches of the Chinese industry sector (17 branches in mining, 165 branches in manufacturing and 7 branches in public utilities). The data source is *The Data of The Third National Industrial Census of The People's Republic of China in 1995*, which is the latest and most comprehensive database published by the SSB in 1997. The scope of data includes all industry enterprises with an independent accounting system.

A few notes about the data should be made before estimating production function. There is a considerable difference between SOEs and FIEs in the proportion of labor and capital that are directly utilized in the production process. In general, Chinese state-owned enterprises allocate a relatively larger proportion of their total capital and labor than FIEs for some non-production purposes such as residential housing, health care, education and other forms of welfare provided to their employees. Consequently, the proportion of inputs that are used directly in the production process in SOEs is generally smaller than that of FIEs.

In order to make an accurate assessment and to improve the estimation efficiency, the data for capital and labor inputs in SOEs and FIEs should be adjusted using the proportion of inputs that are used for production purposes. The proportion of labor and capital used for non-production purposes should be excluded from the production function estimate.

Based on the information provided in the *Third Industrial Census in 1995* (SSB), the ratio of the fixed capital used directly in production to the total fixed capital owned by firms is calculated for both SOEs and FIEs in the 189 industries. As for labor inputs, the separate classification of labor for production and non-production purposes is unavailable in the statistical material. In some previous studies (e.g. Chen *et al*, 1988, P.576 and Reynolds, 1987, P.98), the share of productive fixed capital is used as an approximate value for the share of labor employed directly in the production process. Thus this approach is adopted in this study to measure the share of labor used directly in production.

Another factor worth highlighting is 'working capital', a component of total capital. Working capital is necessary for firms to finance stocks of materials and inventories of finished products and to pay labor, thereby maintaining production. A lack of working capital may limit production expansion. In some previous studies, the data of working capital have been included in the production function along with fixed capital (e.g. Wu, 1995). However, there are two problems associated with inclusion of working capital in the estimation of production functions.

First, the effect of working capital reflects the performance of the distribution system rather than the efficiency of current production. Second, payment to labor is one principal component of working capital. Its inclusion in the production function as a separate variable along with labor and fixed capital will give rise to a biased estimation result, because labor and working capital are highly correlated with each other. Furthermore, in a widely recognized study conducted by the Chinese Academy of Social Sciences, University of Pittsburgh and Brandeis University, the direct estimation of output elasticity with regard to working capital yields a zero coefficient in a time series (1953-1985) study (see Chen *et al*, 1988 for details). For this reason, working capital is excluded from this study.

In this regression estimation, the output (Q) is the value-added of all enterprises with an independent accounting system in each industry. The capital input (K) is the net value of fixed capital on a yearly average, adjusted by its proportion used directly in production. Labor (L) is the number of employees on a yearly average, adjusted by its proportion used directly in production. Since the cross-section data in the given year does not involve a time-series factor, it can be assumed that the technology progress is stable or constant for both state-owned enterprises and foreign-invested enterprises.

5.5.3 Estimation Results and Interpretation

Four models of production functions including the Cobb-Douglas productions function, the CES production function, the Bairam-Vinod and the translog production functions have been estimated. The regression results are presented in Table 5.3 for state-owned enterprises and foreign-invested enterprises respectively.

Table 5.3 Regression Results for Production Function Analysis

A. State-owned Enterprises:

	Cobb-Douglas	Bairam-Vinod	CES	Translog
lnK	0.5084	0.4226	0.5109	-0.8836
	(5.086)*	(4.162)	(2.151)*	(-1.754)
lnL	0.5009	0.4017	0.5141	1.9862
	(4.309)*	(3.405)	(2.104)*	(3.186)*
lnL*lnK		0.0294		0.8481
		(3.111)		(2.644)*
$(lnL-lnK)^2$			-0.1687	
			(-0.671)	
$(lnL-lnK)^3$			0.0784	
			(1.064)	
$(lnL)^2$				-0.5005
				(-2.632)*
$(lnK)^2$				-0.3295
				(-2.429)*
Constant	-1.6428	-1.4434	-1.6281	-2.6288
	(-7.520)*	(6.474)	(-7.443)	(-4.609)
R^2	0.8737	0.8800	0.8751	0.8844
F statistics	986.09	776.50	592.30	532.84

B. Foreign-Invested Enterprises:

	Cobb-Douglas	Bairam-Vinod	CES	Translog
lnK	0.5474	0.5458	0.4184	0.3794
	(8.906)*	(8.763)*	(5.321)*	(4.918)*
lnL	0.5233	0.5211	0.6437	0.7244
	(7.651)*	(7.470)*	(7.695)*	(7.708)*
lnL*lnK		0.0016		0.2460
		(0.1727)		(3.300)*
$(lnL-lnK)^2$			-0.0553	
			(-1.330)	
$(lnL-lnK)^3$			-0.0205	
			(-1.431)	
$(lnL)^2$				-0.1573
				(-3.569)*
$(lnK)^2$				-0.0915
				(-2.668)*
Constant	-1.0664	-1.0718	-1.041	-1.0601
R^2	0.9154	0.9154	0.9192	0.9209
F statistics	838.20	625.38	522.82	443.38

* Statistically significant at 1% level. The sample size is 189.

From these results, we find that the Cobb-Douglas production function is justified as an appropriate form of production function for both state-owned and foreign-invested enterprises. The CES, the Bairam-Vinod and the translog production functions are rejected because the parameters for all interactive items including $\ln K * \ln L$, $\ln L^2$, $\ln K^2$, $(\ln L - \ln K)^2$ and $(\ln L - \ln K)^3$ are not significantly different from zero. In addition, the inclusion of some of these interactive items leads to biased results. For example, the coefficient of fixed capital (K) for state-owned enterprises is negative in the translog production function estimate, indicating an inverse relation between output and investment. This is not economically plausible. Therefore, the CES model, the Bairam-Vinod and the translog models are rejected in favor of the Cobb-Douglas model.

Using the estimated coefficients from the Cobb-Douglas model, the estimated production function for both SOEs and FIEs can be specified as:

For SOEs: $\ln Q = -1.643 + 0.508 \ln FK + 0.501 \ln L$;
For FIEs: $\ln Q = -1.066 + 0.547 \ln FK + 0.523 \ln L$

The estimation results suggest that constant returns to scale prevail in both state-owned enterprises and foreign-invested enterprises in the 189 industrial branches of China. The returns to scale ($\alpha + \beta$) for SOEs, is 1.01, and for FIEs is 1.07. This confirms that a constant returns to scale Cobb-Douglas production function can be justified as the underlying model of Chinese aggregate manufacturing production.

By comparing the estimated production functions for SOEs and FIEs, a few interesting points can be found. First, the returns to scale for FIEs is slightly higher than that for SOEs. This is associated with a relatively greater elasticity of output with regard to labor and capital in FIEs than in SOEs. A one percent increase in fixed capital would give rise to a 0.547 percent growth in output for FIEs, compared to a 0.508 percent growth of output in SOEs. Similarly, a one percent increase in labor results in a 0.523 percent growth of output in FIEs compared to a 0.501 percent growth of output in SOEs. This may be attributed to a difference in technical and management efficiency. A higher technical efficiency and rationalized management are expected to contribute to a higher growth of output for a given increase in inputs.

Second, as the coefficient of the constant term (ln A) in the Cobb-Douglas production function is a measure the level of technology, a comparison of the estimated value of the constant terms (intercept of the production function) for the two types of firms will indicate the difference in technology applied in production between them. It can be seen from Table 5.3, the estimated value of the constant term for FIEs is -1.066, significantly higher than that for SOEs (-1.643). This indicates that the level of technology is higher in FIEs than in SOEs, and suggests that FIEs are more technically efficient compared to SOEs.

Using the elasticity of output with respect to labor and capital obtained from the production function estimation and the average productivity of the two inputs listed in Table 5.2, the marginal productivity of labor and capital for the two types of firms can be calculated. Since output elasticity E_x is defined as

$$E_x = \partial(\ln Y) / \partial (\ln x) = (x/Y) (\partial Y/ \partial x) = MP_x / AP_x$$

where $x = L, K$, the marginal productivity of labor (or capital) is equal to the elasticity of output with regard to labor (or capital) multiplied by the average productivity of labor (or capital). They are: $MP_L = E_L AP_L$; and $MP_K = E_K AP_K$. The marginal productivity of labor (MP_L) and marginal productivity of capital (MP_K) are calculated and presented in Table 5.4.

Table 5.4 Productivity Comparison between FIEs and SOEs

	E_L	E_K	MP_L	MP_K	AP_L	AP_K
SOEs	0.501	0.508	29.06	0.62	58.0	1.21
FIEs	0.523	0.547	84.83	1.12	162.2	2.15

Note: AP_L = output/number of employees; MP_L= increased output produced by using one more employee; AP_K = output / fixed capital, and MP_K= increased output associated with use of one more unit of fixed capital. The unit for AP_L and MP_L is thousand *yuan* and the unit for AP_K and MP_K is *yuan*.
Source: the elasticity of output with regard to labor (E_L) and capital (E_K) are from Table 5.3. and the average productivity of labor (AP_L) and capital (AP_K) are from Table 5.2.

As shown in Table 5.4, foreign-invested enterprises lead state-owned enterprises in both average product and marginal product of labor and capital. We can thus conclude that foreign-invested enterprises in general have higher productivity. Furthermore, the AP_x is larger than MP_x for both

FIEs and SOEs, implying that both of them are producing within the reasonable regions.

In fact, there are many factors contributing to the differences between domestic firms and FIEs in factor productivity. In addition to the factors discussed previously, such as capital intensity, technology content, labor quality and management efficiency, the ratio of the utilization of productive capacity is an important variable affecting the productivity of labor and capital. Due to the transition from a centrally-planned system to a market economy, Chinese domestic firms, especially state-owned enterprises, face various difficulties including financial, marketing and management problems. As a result, the utilization ratio of productive capacity in these firms is much lower than that in FIEs. In most cases, the utilization ratio is lower than 70 percent in SOEs. According to a recent comprehensive investigation undertaken by the State Committee for Regime Reforms in May 1996, 62.3 percent of the large and middle-sized state-owned enterprises in the 12 sampled provinces[2] are run at the capacity utilization ratio of 70 percent or lower, 41.7 percent of the large and middle-sized state-owned enterprises are only operated partly, with capacity being utilized at 50 percent or lower.

A lower utilization ratio of capacity results in lower productivity of factors and a considerable waste of economic resources. According to a recent statistics published by the SSB, 33.8 percent of SOEs ran at loss in 1995 (SSB, *The Third National Industrial Census 1995*). This situation appears to have deteriorated in 1997. The SSB officers indicated that SOEs as whole ran at loss in the first half of 1997 in China.

In addition, export behavior is also an important explanatory factor. As listed in Table 5.1, the export-output ratio of FIEs is remarkably higher than that of domestic firms. For example, in 1991,1992 and 1995, the export-output ratio for FIEs was about 40 percent compared to 10 percent for domestic firms. In general, export expansion helps to increase productivity of factors through its positive effects on the efficient allocation of resources and productive and management efficiencies.

Finally, the structure of property ownership also plays an important role in determining different productive efficiency between SOEs and FIEs. Under the current state-ownership regime, capital and labor are allocated by the state plans rather than by market mechanisms. The prices of output and inputs are frequently distorted and deviate from their market values. Therefore, allocation and utilization of resources including investments and labor are in many cases not economically rational. Furthermore, since the government assumes responsibilities for the profits and losses of SOEs, there is no direct linkage between enterprises' profitability performance and their employees' benefits. As a result, employees, including managers and workers in these enterprises, have less incentives than their counterparts in FIEs to improve productive efficiency.

Accordingly, profit-maximization is not the major goal for many state-owned enterprises, which rather tend to target maximization of employment and workers' welfare. In the current economic transition from a planing system to market-oriented economy, these enterprises face a dilemma: a conflict between profit maximization and employment maximization. To achieve the maximum profit, state-owned enterprises need to rationalize their operations which necessarily means cuts in employment. This is likely to produce some social and political problems. On the other hand, to maintain the employment level, the goal of profit maximization has to be sacrificed. This dilemma slows progress in reforms of state-owned enterprises, especially in ownership transformation. This is a fundamental reason why the productivity of state-owned firms are persistently lower than that of FIEs.

5.5.4 A Discussion and Comparison with Other Similar Studies

Several studies have been undertaken to compare the productivity and efficiency of foreign firms and domestic firms in various countries. In the case of Iran, Asheghian (1982) found that Iranian-American joint venture firms were more efficient than local Iranian firms in terms of labor productivity, capital productivity and total factor productivity. In his comparative study of foreign and local Brazilian firms, Willmore (1986) found that foreign firms have a higher ratio of value-added to output, a greater export-output ratio, higher labor productivity and wage rates, and a greater capital intensity. These findings are substantially consistent with the results of this study.

Unlike Sterner's (1990) study of the Mexican cement industry, which found that multinationality has no statistically significant effect on efficiency, this study finds that foreign involvement in capital and technology has positive impact on efficiency. Productivity performance of foreign-invested enterprises in China is superior to that of domestic firms, especially the state-owned enterprises. Similarly, the findings of this study also differ from those by Schive (1982) and Tu (1990), who found that foreign firms in Taiwan tended to use more labor-intensive production method than domestic firms. This study finds that foreign-invested enterprises tend to use more capital and less labor compared to domestic firms. In addition, Tu (1990) found that the marginal productivity of capital in foreign firms was higher than that of domestic firms, whereas the marginal productivity of labor was lower in foreign firms than in domestic firms. This study finds that marginal productivity for both labor and capital in FIEs are higher than that of state-owned enterprises.

In terms of methodology applied in estimation, this study has tested four models of production functions including the Cobb-Douglas, the CES, the Bairam-Vinod and the translog production functions. Based on the results from each estimation, the Cobb-Douglas production function is justified as an appropriate specification of production function for both FIEs and SOEs. This is consistent with Bairam's findings from his study of ex-Soviet Union (1987) and his study of New Zealand manufacturing industries (1994).

To conclude this section, a note of caution is necessary in interpreting the estimation results. First, most FIEs have been operating in China for a short period, normally five or six years. Therefore, their production behavior and characteristics have not been fully demonstrated, hence data on their operation can only reflect their production behavior and efficiency at the first stage. Second, the data used in this estimation is cross-section data (189 industry branches) in a given year (1995); the estimation results therefore cannot directly provide an indication of technical progress over time and its effects on productivity. Third, due to lack of reliable data on prices of labor and capital, the cost functions are not estimated in the study. The findings of this study are therefore indicative rather than conclusive.

5.6   Conclusions

This chapter has presented a comparative study of the characteristics and productive efficiency of foreign-invested enterprises and domestic firms. It has demonstrated that in general FIEs, compared to domestic firms, have a higher capital intensity, use capital-intensive production methods, and have a greater export-output ratio. In addition, FIEs have a higher incremental output-capital ratio (IOCR), and exhibit higher productivity for both capital and labor

The higher productive efficiency of FIEs can be attributed to many factors, such as higher capital and technology intensities, modern management expertise, better trained employees and efficient organization of the production process. In addition, some other factors have an important bearing on the efficiency difference between FIEs and domestic firms. The ownership structure of enterprises is a critical factor affecting the productive efficiency of different types of firms. The findings of this study suggest that state-ownership tends to affect the productive efficiency of enterprises negatively.

The limitations of this study need mentioning. Due to the paucity of time-series data over a reasonably long period, the cross-section data for the year 1995 was used in the production function analysis. As a result, technical progress and its impact over time are unable to be measured in this estimation. In addition, only partial factor productivity (i.e. labor productivity and capital productivity) is estimated in this study. Total factor productivity for FIEs and domestic firms is not estimated due to lack of reliable data on factor prices especially the real wages information.

Notes

[1] In Chinese statistics, the industry sector include mining, manufacturing and public utilities such as power and water production and supply. The manufacturing industry is the core of the industry sector. In 1995, the manufacturing industry accounted for 89.6 percent of the total value of output in the industry sector, while the mining and public utilities industries accounted for 6.2 percent and 4 percent respectively (SSB, 1997).

[2] The 12 provinces are Shanghai, Tianjin, Liaoning, Shichuan, Guangdong, Zhejiang, Jilin, Henan, Hunan, Shanshi, Guizhou and Yunnan.

6 Industrial Linkage Effects of DFI on Domestic Sectors

6.1 Introduction

DFI affects the host economy both directly and indirectly. It directly affects domestic capital formation, economic growth, exports, balance of payment, employment, technology transfer and government tax revenue. It also influences domestic sectors through industrial linkages. In order to assess the overall impacts of DFI on the Chinese economy, it is necessary to study the linkage effects of DFI on the domestic sector of China.

This chapter firstly presents a theoretical framework for an analysis of linkage effects, and then probes the potential linkage effects of DFI in the Chinese industry sector by estimating industrial linkage indices and discussing the industrial distribution of DFI activities. An investigation of the extent to which potential linkage effects have been realized is exemplified with case studies of the automobile and electronics industries.

6.2 Theoretical Framework

The linkage analysis is based on an input-output table. In the past 30 years, a large number of studies have contributed to the literature of linkage analysis. Leontief was the first to formulate the theory of inter-industrial linkage. Following his tradition, Rasmussen (1956) and Hirschman (1958) did pioneering works in developing the linkage theory. Since then, many studies have been conducted (e.g. Chenery, 1961; Little and Gerald, 1968; Hazari, 1970; Yotopoulos and Nugent, 1973; Schultz, 1976; McGilvary, 1977; Bulmer-Thomas, 1982; Heimeler, 1991; Clements and Rossi, 1991). Some authors have explored the linkage effects or the spillover efficiency of foreign investment in host countries. For instance, Blomstrom and Hakan (1983), and Blomstrom (1989) explored the spillover efficiency of foreign investment in the case of the Mexican manufacturing industry. O'hUallachain (1984) examined the input-output linkage effects of DFI in Ireland. Recently, Schive (1990) investigated the linkage effects of DFI in the case of Taiwan. Using panel data for Morocco, Haddad and Harrison (1993) examined the spillover effects of foreign firms in the manufacturing sector.

To date, no attempt has been made to study this issue in the case of China. This chapter aims to fill in this gap by investigating the potential linkage effects of DFI and their materialization in China's industry sector. Before empirical investigation is undertaken, a discussion on the theoretical framework is outlined.

6.2.1 Linkage Indices

Input-output linkages can be classified into two types, viz. backward and forward linkages. A backward linkage refers to a derived input demand effect. It measures the increase in production of all other industries generated by an increase in final demand for the products of a given industry. Backward linkage effects arise if one non-primary sector activity induces supply of the inputs needed in that activity through domestic production (Hirschman, 1958, p.100). Forward linkage, on the other hand, represents the demand of every other sector for the products of a given sector. The forward linkage indices measure the degree of response of the output of other sectors to a larger input supply provided by a given sector. High forward linkage effects occur when a sector's output is, or could be used, as an input by many other sectors. In other words, the forward linkage represents the linkage effect induced by output supply, while backward linkage measures the effect created by input demand.

In industrial linkage analysis, inputs refer to machinery and intermediate inputs such as raw materials, semi-finished products, spare parts and fittings. Through a certain production process these intermediate inputs can be transformed into products, which can be used as either consumer goods or inputs for production of other goods. Labor is excluded from the input-output analysis. This is because labor is not a product of any particular industry although it is an essential input for all kinds of production process.

In essence, linkages refer to the inter-sectorial relations or potential inter-sectorial impacts within an economy. Many factors may influence the inter-sectorial relations and the materialization of potential linkage effects. These factors include the nature and relative prices of products, the development of markets and government policies. There is no guarantee that the potential stimulus measured by backward and forward linkages will be translated into actual growth.

In general, manufacturing industries have higher backward linkage effects than primary industries since they use other sectors' products as inputs for further processing. Primary industries such as agriculture, fishing, forestry and mining industries have lower backward linkage effects as they use natural resources rather than man-made inputs for their production. Since their products normally need further processing in other sectors, these primary industries possess relatively high forward linkage effects. Similarly, industries that produce high value-added products have higher backward linkage effects than those producing low value-added products. In addition, well-functioning and properly-developed markets (for labor, products, finance) and appropriate industrial policies tend to facilitate the realization of the potential linkage effects.

The linkage indices of an industry can be measured from an input-output table. Applying the Leontief inverse matrix, Rasmussen (1956) initially normalized the measurement procedures of linkage effects. Following his procedures, Yotopoulos and Nugent (1973), McGilvary (1977), Bulmer-Thomas (1982), Schive & Majumdar (1990), West (1993) and others, measured linkage effects by comparing the 'average' stimulus created by, and in response to, changes in final demand in a sector with the overall average. The mathematical formula to measure backward linkage is specified as:

$$Y_j = \frac{1/n \sum_i r_{ij}}{1/n^2 \sum_i \sum_j r_{ij}} = \frac{\sum_i r_{ij}}{1/n \sum_{ij} r_{ij}}$$

where r_{ij} denotes the ith row and jth column element of the Leontief inverse matrix $(I-A)^{-1}$, and n is the number of sectors. The numerator is the column summation, referring to the total increase in the output of the economy to be absorbed by, or supplied to, the jth sector for its one unit increase in final demand. It denotes the average stimulus imparted to other sectors by one unit change in demand for sector j. The denominator denotes the average stimulus for the economy as a whole when demand increases by unity. If $Y_j > 1$, the increase in demand in jth sector will yield above average backward linkage effects, while the opposite is true if $Y_j < 1$. (Bulmer-Thomas, 1982).

The formula for forward linkage index is

$$Y_j = \frac{1/n \sum_j r_{ij}}{1/n^2 \sum_i \sum_j r_{ij}} = \frac{\sum_j r_{ij}}{1/n \sum_{ij} r_{ij}}$$

If $Y_i > 1$, the ith sector will be more heavily drawn upon by the expansion of the economy, i.e. the output of sector i is broadly and intensively used as inputs by other sectors (user sectors). An increase in the output of sector i could induce user sectors to expand output, as user sectors take advantage of the increased availability of inputs. If $Yi < 1$, it indicates that the products of sector i are primarily used for final consumption, rather than used as inputs for production of other goods.

Another concept associated with linkage indices, is the coefficient of variation (the spread indices), which measures the variation in the distribution of the linkage effects among all sectors. As the backward linkage is defined as the column mean divided by the average of the column means in the Leontief open inverse matrix, the backward spread index is defined as the column coefficient of variation divided by the average coefficient of variation over all sectors. Likewise, forward linkage is defined as the row mean divided by the average of the row means for all sectors, while the forward spread index is defined as the row coefficient of variation divided by the average coefficient of variation over all sectors (West, 1993). A lower variation coefficient implies the linkage effects are distributed evenly in all sectors, whereas a higher variation coefficient indicates the linkage effects are concentrated in a few sectors.

6.2.2 Multipliers

A multiplier coefficient measures the increase in the total output of the economy in response to one unit of the initial increase in final demand in an industry. If a one dollar initial increase in the final demand of a given industry induces a three dollars increase in the total output of the economy, then the output multiplier is three. In mathematical terms, the output multiplier is the column sum of the Leontief inverse, and can be expressed in the following formula:

$$M_j = \sum_i r_{ij}$$

where M_j is the output multiplier for industry j and r_{ij} denotes the $(i, j)^{th}$ element of the Leontief inverse matrix. Thus, the multiplier is equal to the numerator of backward linkage formula. Hence, it provides a measure of backward linkages, which traces the cumulative effects of a unit increase in the final demand of a specific sector on the input requirements from all sectors. A high backward linkage index for a sector is expected to generate a large output multiplier. Analogously, a large multiplier suggests the existence of a high backward linkage effect of a sector.

The initial increase in final demand can be in any category of final demand, e.g. private consumption, investment, government expenditure and exports. Through backward linkage effects, the initial increase in final demand will induce the second and subsequent round impulses which pass through the economy. As a result, the overall output (also employment and income) of the economy will increase by a multiplier.

As mentioned previously, the linkage indices and multiplier are potential effects existing in an economy owing to the industrial interdependencies in input and output. To what extent can the potential linkage effects be materialized? This depends on the economic structure of a country. In a closed economy, firms (or sectors) depend on each other for both input supply and output markets. The inputs for every sectors are acquired from other domestic producers; and, conversely, their outputs are supplied to each other. As a result of these linkages, an initial investment in one sector will induce a series of investments in other sectors, and thereby giving rise to more employment and output throughout the economy.

In an open economy, however, the linkage effects may be different. On the one hand, firms can import inputs from overseas and export outputs. In terms of backward linkage, the importation of inputs may result in a *leakage* of final demand, and therefore weaken the linkage effects on the domestic economy. In extreme cases, if the inputs used by foreign firms in a host country are totally imported and all output is exported, a foreign enclave will form. In this situation, foreign firms have little industrial linkage with local firms. On the other hand, if foreign firms largely utilize the local inputs and export some of their output, the backward linkages will be quite high. Local firms in the host country will benefit from such linkage effects by supplying inputs used for the production of exports. Therefore, the *local content of inputs* used in production by foreign firms is an important indicator measuring the linkage effects of foreign firms.

In linkage analysis and industrial planning, backward linkage effects are usually given more attention. This is because backward linkage indices measure the effects of a sector on the whole economy from the demand side and identify the stimulus created and passed by a sector on to the economy. Thus, they enable policy makers to choose *key industries* (which is defined in the input-output analysis as the industries with high backward linkage indices) to accelerate development.

In this chapter, the discussion focuses on backward linkage effects of foreign-invested enterprises (FIEs) on the domestic sector of China. Specifically, it investigates (1) the potential linkage effects of FIEs on the Chinese economy, and (2) the extent to which these potential linkage effects have been materialized in different sectors. The first issue can be explored by examining the backward linkage indices of economic sectors and the industrial distribution of DFI. If DFI or FIEs' production activities are concentrated in the industries with high backward linkage indices, their potential linkage effects are expected to be high. The second issue will be explored by investigating the local content of intermediate inputs used in FIEs.

6.3 The Potential Linkage Effects of DFI in China

The backward linkage indices can be calculated by applying Rasmussen's formula to the input-output table. In principle, this index shows the share of an economy's gross production which is absorbed by, or supplied to, a certain industry in response to its one-unit increase in final demand (Schive, 1990, p.69). Using the latest version of *the Input-Output Table of China* (1992) and *the IO7 computer program* developed by West (1993), the backward linkage indices and output multipliers have been calculated for 34 sectors of the Chinese economy (see Appendices 6.1 and 6.2). Based on these indices, all these industries can be ranked, and hence the 'key industries', which are industries with high linkage indices and multipliers, can be identified.

Table 6.1 presents the backward linkage indices and multipliers for 22 industry branches which compose the main body of the *industry sector*[1] in Chinese statistics. Of the 22 industries, 15 industries present backward linkage indices exceeding unity. Accordingly, the output multipliers in all 15 industries are larger than three. Therefore, these 15 industries can be identified as 'key industries' of the Chinese economy in terms of industrial linkage and multiplier effects.

Table 6.1 Backward Linkage Indices, Multipliers and DFI

Industries	Linkage Indices	Rank	Output Multiplier	DFI[1]	%	FIE Output[2]	%[3]	FIEs' % in total output
Clothing, shoes	1.206	1	3.587	249.2	8.0	1259.8	11.8	51.5
Electronics	1.187	2	3.597	302.9	9.6	1517.7	14.2	64.6
Textile	1.182	3	3.545	250.9	8.0	824.1	7.7	17.9
Metal Products	1.175	4	3.570	155.6	4.9	439.4	4.1	26.6
Electrical Products	1.165	5	3.522	184.3	5.9	631.4	5.9	24.3
Transport Equipment	1.164	6	3.510	163.1	5.2	813.4	7.6	24.6
Miscellaneous goods	1.163	7	3.517	372.8	11.8	1006.1	9.4	22.2
Wood Products	1.147	8	3.484	61.9	2.0	184.1	1.7	29.2
Machinery	1.133	9	3.449	165.3	5.3	491.4	4.6	11.9
Paper & Printing	1.109	10	3.353	176.1	5.6	432.7	4.0	25.8
Coal Products	1.098	11	3.603	0.7	0.01	5.6	0.05	6.3
Metal smelting	1.094	12	3.417	79.8	2.5	403.4	3.8	8.0
Chemicals	1.091	13	3.380	158.1	5.0	503.5	5.0	13.2
Precise Instruments	1.070	14	3.252	54.3	1.7	168.8	1.6	39.6
Building Materials	1.021	15	3.182	223.5	7.1	352.5	3.3	11.7
Food & Beverage	0.984	16	2.892	357.5	11.3	1194.7	11.2	23.0
Metal mining	0.967	17	3.158	1.3	0.0	2.4	0.0	0.05
Petroleum processing	0.949	18	3.138	3.9	0.1	23.3	0.2	1.2
Coal mining	0.934	19	3.133	0.6	0.02	2.8	0.01	0.2
Non-metal mining	0.923	20	2.941	7.4	0.2	12.5	0.1	3.4
Power and steam	0.857	21	3.048	173.7	5.5	336.7	3.1	13.8
Oil, gas exploring	0.755	22	2.753	0.2	0.0	57.8	0.5	4.1
All Industries	1.000			3148	100	10714	100.	19.5
Rank correlation[4]							0.57	0.62

Notes: 1. DFI figures refer to the actually used DFI in registered FIEs in 1995. 2. FIEs' output is for 1995. The unit for DFI and FIEs' output are 100 million *yuan*. 3. The percentage in columns 6 and 8 are the distribution percent of DFI and FIEs output in each industry respectively. The percentage in column 9 is FIEs' output as a share in the total output of each industry. 4. The rank correlation refers the correlation between the rank of backward linkage indices and the rank of the distribution percentage of FIEs' output in each industries, and FIEs' share in the total output value of each industry.

Source: SSB, *The Third National Industrial Census of the People's Republic of China.*

In order to investigate the potential backward linkage effects of DFI on Chinese domestic sectors, it is necessary to examine (1) the industrial distribution of DFI and FIEs production, and ascertain whether DFI and hence FIEs' activities are concentrated in the industries with high backward linkage indices; (2) explore FIEs' shares in the total output of each industry, measuring the significance of FIEs in each industry; and (3) determine the rank correlation coefficients between industries' linkage indices and the industrial distribution share of DFI and FIEs output.

As shown in Table 6.1, DFI and FIEs' output are highly concentrated in the industries with high backward linkage indices and output multipliers. DFI in the 15 'key industries', whose backward linkage indices are larger than one, accounted for 83 percent of the total DFI in the industry sector in 1995. Similarly, 85 percent of total FIEs' output was produced in these 15 industries. This pattern of distribution of DFI and FIEs' production indicates that FIEs have high potential linkage effects on Chinese domestic sectors. FIEs may impart strong potential linkage effects to domestic firms and stimulate their growth by providing additional demand.

In terms of industry significance, FIEs have become important players in Chinese industrial production. Overall, FIEs produced 20 percent of total output of the industry sector in 1995. In some industry branches, the role of FIEs is even more significant. For example, 64.6 percent of electronic products, 51.5 percent of clothing, shoes and leather products, and 39.6 percent of precise instruments in China were produced by FIEs in 1995. In other industries, FIEs also are important producers. As can be seen in Table 6.1, with the exception of the food and beverage industry, all the industries in which FIEs' production activities are concentrated present high backward linkage indices. As important producers in Chinese industry sector, therefore,, FIEs can promote the growth of domestic sectors through industrial linkage and multiplier effects.

In addition, the high potential backward linkage effects of FIEs on domestic sectors can also be examined by testing rank correlation coefficients between the backward linkage indices of industries and the distribution percentages of DFI and FIEs' activities. As shown in Table 6.1, the rank correlation coefficients between the backward linkage indices and the distribution percentage of FIEs' output is 0.56, the rank correlation coefficient between the backward linkage indices and FIEs' share in the total output of each industry is 0.62. Therefore, the positive correlation between the distribution of FIEs' production activities and the backward linkage indices further confirms the potential linkage effects of FIEs.

6.4   Materialization of the Linkage Effects of DFI: Case Studies

Since FIEs' backward linkage effects are materialized through use of locally made (as opposed to imported) inputs in the production process (Schive and Majumdar, 1990), it is necessary to investigate the local content of inputs used by FIEs in production. By examining the share of locally made intermediate inputs in total inputs used by FIEs, an estimate can be made of the extent to which the potential backward linkage effects of FIEs have been realized. Since the local content of inputs used in FIEs is the ratio of locally made materials and parts to the total intermediate inputs used in FIEs, it can be calculated for each industry if imported inputs are known, as locally made inputs equal total intermediate inputs less imported inputs.

However, due to paucity of systematic statistical data on imported inputs used in FIEs at the industry level, it is extremely difficult to precisely measure the imported inputs in FIEs for an industry. This is because some goods imported by FIEs are sold to domestic enterprises, rather than used in their own production process. Further, the available official statistics of foreign trade by FIEs are classified by commodity category rather than by 'inputs' or 'final products'. Although some imports can be identified as 'inputs', e.g. wool and minerals, a large number of imported commodities are difficult to identify as 'inputs' or 'final products'. With these statistical limitations, the local content of inputs used in FIEs may be evaluated using surveyed data at the firm level.

This section probes the extent to which the potential backward linkage effects have been realized, using case studies of two industries, i.e. the automobile industry and the electronics industry. These two industries are characterized by high backward linkage indices. As depicted in Table 6.1, the electronics industry and the transport equipment industry (automobile fall into this category) are ranked respectively the second and the sixth industry with the highest backward linkage indices. Further, they are fast-growing industries in China, with increasing shares in both total output and DFI in the manufacturing sector. The data used in this study are acquired from Chinese government departments.[2]

6.4.1   *Local Content of Inputs Used by FIEs in the Automobile Industry*

The automobile industry is a newly established industry in China which has grown rapidly in the last 5 years. In 1995, the industry produced 3.29 million automobiles, with a sale revenue of 174 billion *yuan*, accounting for 3 percent of Chinese GDP. Foreign-invested enterprises played an important role in this development, with more than 549 FIEs operating in this industry by 1995, there. FIEs help upgrade the technology and product quality of local firms through joint venture mechanism and linkage effects, contributing greatly to the industry development, In 1992 alone, over 200 technology transfer projects were introduced by FIEs to Chinese automobile producers (Ministry of Machinery Industry, 1994).

Localization of inputs is an important feature of the automobile industry development. Since China's automobile market is huge and characterized by a chronic excess of demand in relation to supply, the domestic prices of automobiles have remained quite high, normally two or three times as much as international market prices. For this reason, many giant car manufacturers from the U.S. Japan and Europe target this market. The depreciation of the *Renminbi* (the Chinese currency) against major hard currencies in recent years has resulted in imported parts and fittings becoming more expensive in terms of *Renminbi*. Car manufacturers are now motivated to use locally-made parts and fittings from domestic firms. This localization campaign has been further supported by Chinese government policies such as the imposition of high tariffs on imported intermediate inputs that are not used for production of exports.

The car manufacturing industry is a good example of the automobile industry development. The annual output of this industry grew from 40 thousand cars in 1990 to 519 thousand cars in 1995. The bulk of this output was produced by the following five FIEs: Shanghai Volkswagen AG, Beijing Jeep Corporation, Guangzhou Peugeot Automobile Corporation, the First Auto-Volkwagen Corporation (in Changchun) and Tianjin Daihatsu Corporation. The first three corporations are the largest FIEs in China's industry sector. The First Auto-Volkwagen Corporation was the 16th largest FIE in China's industry sector . Tianjin Daihatsu Corporation was also listed in the top 30 largest FIEs in the industry sector of China in 1995. In 1995, these five FIEs produced about 90 percent of the total output of the Chinese car manufacturing industry.[3]

All five car manufacturing FIEs are Sino-foreign joint ventures. They have made remarkable progress in the localization of parts and fittings used in car manufacturing, with the local content of inputs used by these FIEs increasing significantly in the past few years, as shown in Table 6.2. As the local content of inputs has increased, the backward linkage effects of these car producers have been increasingly materialized. Therefore, local firms in this industry and many other associated industries are stimulated by FIEs' increased demand for locally made intermediate inputs. This can be illustrated by the cases of Shanghai Volkswagen Corporation and Beijing Jeep Corporation.

Table 6.2   The Local Content[1] of Parts and Fittings in Five Automobile FIEs (%)

Name of FIEs	1988	1989	1990	1991	1993
Shanghai Volkswagen	12.6	31.0	60.0	75.3	80.5
Beijing Jeep	30.0	35.1	44.0	57.3	60.5
Tianjin Daihatsu	12.5	40.0	40.7	47.4	61.8
Guangzhou Peugeot	n/a	27.6	n/a	51.5	n/a
The First Auto-Volkswagen	n/a	6.6	6.7	30.7	44.0

Note: 1. The local content percentage refers to the proportion of locally made materials and intermediate products in the total intermediate inputs used in FIEs. The intermediate inputs are materials, engine, parts and fittings. Labor input and fixed capital input are not included.

Source: *Quanguo Qiche Gongye Fazhan Huiyi Zilao Huibian (The Proceedings of the National Conference on Automotive Industry Development)* 1993, P.64, this is an unpublished official document. *Zhongguo Xinxi Bao (the Information Daily of China)*, May 13, 1994.

Shanghai Volkswagen Corporation is a Sino-German joint venture established in 1985, producing the Shanghai Xantna model. Since the early stages of operation, this company has been highly motivated to localize parts and fittings used in car production. The local content of parts and fittings was only 5.71 percent in 1987. By 1993, it had risen to 80.47 percent. Xantna car production involved 1592 spare parts bought locally from more than 200 factories across 10 industries. As the localization progressed, the output of this company grew rapidly from 18,000 cars in 1990 to 100 thousand cars in 1993. The output produced by this company accounted for 40 percent of the national total.[4] A large production network led by this company has been formed, with many local factories being involved either horizontally or vertically in car manufacturing. Therefore, it is clear that the production expansion of Xantna car would stimulate the development of many local factories through backward linkage effects.

Another example of backward linkage effects is provided by Beijing Jeep Corporation. This is a Sino-U.S. joint venture established in 1984 by

Beijing Automobile Manufacturing Company and American Automobile Company. The major product is the Beijing Charade Jeep. In the past ten years this company has made great efforts to localize the parts and fittings used in car manufacturing. By 1993, 167 enterprises across 21 provinces were involved in its 'localization community'. Its local content of inputs rose sharply from 1.73 percent in 1985 to 60.48 percent (Beijing Jeep Corporation, 1994). In addition, this company put equal emphasis on upgrading its technology and equipment. From 1984 to 1990 it introduced 1587 items of new technology from foreign companies, especially from the American partner company. As a result, the overall annual labor productivity increased from 70,843 *yuan* per worker in 1984 to 227,366 *yuan* per worker in 1990.

Other FIEs in the car manufacturing industry have also experienced an impressive progress in the localization of inputs, as Table 6.2 shows. Many local firms are increasingly involved in FIEs' production processes through backward linkages. To ensure local firms produce good quality parts, the FIEs frequently transfer new technologies to them, including processing technologies. The FIEs also provide technical training services to managers, engineers and other workers in local supplying firms. In addition, the quality requirements by the FIEs for locally made parts are expected to positively affect the product quality of local firms. Obviously, the FIEs contribute significantly to the production expansion and technology upgrading of local firms through input-output linkage effects and other technical and economic relations. As technology diffusions occur, the productive efficiency of local firms can improve. Therefore, as FIEs are increasingly integrated with the local economy, their backward linkage effects are materialized.

6.4.2 Local Content of Inputs Used by FIEs In the Electronics Industry

The electronics industry has also witnessed a rapid growth. During the period 1987 to 1995, the output value of this industry increased by 22 percent per annum. The high growth of the electronics industry is attributed partially to the rapid expansion of DFI. By 1995, there were 2900 FIEs operating in the electronics industry, producing nearly 65 percent of the total output of this industry (SSB, 1997). There were over a half million Chinese employees working for these FIEs in the industry. Since the electronics industry has the second highest backward linkage index and the largest share of FIEs' output, a discussion of the local content of inputs used by FIEs in this industry is important for assessment FIEs' linkage effects.

To investigate the localization of inputs used in FIEs, the author surveyed the top ten largest FIEs in the electronics industry. Most data are from Chinese official reports published in newspapers, other data are based on direct interviews with FIEs' executives during the author's fieldwork in China in 1995. These ten FIEs and their input localization ratios are shown in Table 6.3.

Table 6.3   Input Localization of the Top Ten FIEs in the Electronics Industry in 1993

Name of FIEs	Location	Investing Country	Sales[1]	Input Local Content %
Huaqiang Sanyo Electronics	Shenzhen	Japan	1402.7	64.2
Shanghai Bell Telephone Equipment Co.	Shanghai	Belgium	1317.7	60.0
Shenzhen Konka Electronics Co.	Shenzhen	Hong Kong	1212.8	65.5
Beijing National TV Picture Tubes Co.	Beijing	Japan	1141.2	74.7
Fujian Hitachi TV Sets Co.	Fujian	Japan	848.5	85.0
China Xunda Elevator Co.	Shanghai	Germany	578.5	90.0
Tianjin OTIS Elevator Co.[2]	Tianjin	US	560.0	76.0
Huizhou TCL Communication Equipment Co.	Guangdong	HongKong	381.3	58.5
Jiangxi Ganxin TV Co.	Jiangxi	Japan	300.7	82.0
Motorola Electronics Co.	Tianjin	U.S.	73.5	35.5
The Above Ten Corporation		N/A	7817.0	70.1[3]

Notes: 1. The sales value of these FIEs refers to year 1992, and the unit is million *yuan*. 2. The local content of inputs for standardized products in this company was 95%, while the localization ratio for newly developed products was only 60%. The average ratio was 76%. 3. The localization ratio of inputs for the sampled ten FIEs is the weighted average ratio, the weights are the share of each company in the sum of their total sales.

Source: *China Economic News*, December 13, 1993; *Zhongguo Dianzi Bao (China Electronics Daly)*, January 6, 1995, January 17, 1995, February 14, 1995; *Economic Daily*, July 20, 1995; and the telephone interviews with FIEs located in Shenzhen by the author in July 1995.

Since the ten largest FIEs are invested by foreign firms with a wide range of nationalities (including Hong Kong, Japan, the US and European countries), and are located in different regions, they are appropriate as the samples of FIEs for study of input localization in the electronics industry. In past few years, these FIEs have made a significant progress in localizing inputs. Locally made materials, parts, components and fittings accounted for an increasing share in their total intermediate inputs, and gradually replaced for imported inputs.

As Table 6.3 shows, the weighted average local content (localization ratio) for the ten FIEs in this industry was about 70 percent in 1993. Of these FIEs, television sets producers have achieved a relatively higher ratio of local inputs. For example, in Fujian Hitachi TV Sets Corporation, which is one of the top five TV producers in China, 85 percent of the total intermediate inputs including parts, components and cells were locally made. In Jiangxi Ganxin TV Corporation, locally made intermediate inputs accounted for 82 percent of the total inputs. Through technology transfer and inputs localization progress, FIEs played an important role in promoting the development of domestic firms in the television manufacturing industry. As a result, the production capacity and product quality of this industry have remarkably improved. By 1994, the localization ratio of intermediate inputs for color television sets with medium and small sized screen was 95 percent.[5]

The performance of input localization in the elevator (lift) manufacturing industry is also impressive. In the two FIEs sampled, the local content of total intermediate inputs reached a high level, 90 percent in China Xunda Elevator Corporation (in Shanghai) and 76 percent in Tianjin OTIS Elevator Corporation. In the later case, the local content of inputs for technically standardized products was 95 percent, for newly developed products it was 60 percent.

In comparison with television and elevators, other electronic products including communication equipment and computer products have relatively lower local content in total intermediate inputs. On average, about 60 percent of inputs used in the production of these technically complex products are locally made. This indicates that domestic firms' capacity for development of technologically advanced products is still at developmental stage and needs to improve. It may also be that foreign investing companies tend to tightly control newly developed technology and are reluctant to transfer new technology to local partner(s). In many cases they are motivated to use imported parts, components and cells from their own affiliates in the home countries or other countries, because by doing that they gain additional benefits. On the one hand, importation of inputs through FIEs expands their exports to China. On the other hand, it facilitates foreign investors' manipulation of transfer prices by over-invoicing imported materials, parts and components, thereby shifting profits to their home countries.

It is not unusual for FIEs to purchase materials and parts from overseas at higher prices than in Chinese domestic markets where these materials and parts are available. Of the ten FIEs listed in Table 6.3, Motorola (China) Electronics Corporation is an example. This company is a wholly-owned subsidiary by two American companies (Motorola and IBM) in Tianjin City. It produces communication equipment (mobile phones) and targets the Chinese domestic market. In 1993, about 65 percent of parts and components used in production were imported from the mother companies in the USA. or other affiliates in their network. Domestically made parts and components accounted for only 35 percent. This is much lower percentage than the local content of inputs used in Shanghai Bell Communication Corporation (60 percent), the largest of the FIEs in the communication equipment manufacturing industry.

The above discussion of the local content of inputs used in FIEs in the automobile and electronics industries suggests that FIEs have made significant progress in localization of inputs in the past few years. In these two industries, the local content of intermediate inputs used in FIEs is about 70 percent on average, with a range from 35 percent to 95 percent. This indicates that the potential backward linkage effect of FIEs in these industries have been materialized to a relatively high extent. In other manufacturing industries especially labor-intensive industries using standardized technology, such as textile, clothing, shoes, toys, food and furniture, the local content of inputs used in FIEs should be higher than that of technology-intensive industries such as automobile and electronic industries.

Since most FIEs have been operating for only a few years, their input-output linkages with domestic firms still have considerable scope to grow and input localization in FIEs can be expected to continue. As FIEs expand, their demand-pull effects (i.e. backward linkage effects) on domestic firms will be increasingly realized, providing strong stimulus to domestic firms with resultant output and employment increases.

6.5 Conclusions and Policy Implications

This chapter has focused on the backward linkage effects of FIEs, and the extent to which the potential linkages effects have actually been realized. It has found that the activities of FIEs are concentrated in industries with high backward linkage indices, with the potential linkage effects of FIEs being quite high. Moreover, the investigation of the local content of inputs used by FIEs in the automobile and electronics industries, indicates that locally made parts and materials constitute the major share of intermediate inputs used in FIEs. As a result, the potential backward linkage effects of FIEs on the domestic economy have been positive. Contrasting to some views, FIEs have not formed foreign enclaves in the Chinese economy, but instead have increasingly integrated with the domestic sectors.

Some policy implications can be drawn. First, an appropriate industry policy is essential for utilization of foreign investment. For a host government, it is important to identify the industries that present high backward linkage effects, and promote foreign investment in these industries. The industrial distribution of DFI should be consistent with the industrial development orientation. The government of a country can use policy tools such as joint venture encouragement, local content requirement, tariff and tax rates, to encourage input localization progress. As FIEs are increasingly integrated into the domestic economy, their linkage effects will become more important.

Finally, legal and economic barriers isolating DFI-concentrated industries and regions from the rest of the economy should increasingly be removed in order to avoid formation of artificially segmented markets and to facilitate the development of nationally united markets. This is a necessary condition to ensure FIEs and local firms connect and integrate with each other.

Notes

1. The 'industry' sector in Chinese statistics includes manufacturing, mining and public utilities such as power production and water supply.
2. They include the Ministry of Machinery Industry, the Ministry of Electronic Industry, Ministry of Foreign Trade and Economic Co-operations and the State Statistical Bureau.
3. Cited from *China Economic News,* July, 13, 1996.
4. *Zhongguo Xinxi Bao*, May 13, 1994.
5. Cited from *Zhongguo Huagong Bao, China Chemical Engineering Daly*, December 12, 1994.

7 Foreign Investment and Regional Economic Growth

7.1 Introduction

Since opening up its economy to the outside world in 1979, China has achieved a remarkable growth. Her gross domestic product (GDP) grew at an average 9.9 percent per annum from 1979 to 1996 (SSB, 1997). However, China's economic growth has been regionally uneven and characterized by increased inter-regional economic disparity and income inequality especially between the *Eastern* (coastal) and *Western* (far inland) regions.[1] The imbalance in economic growth between the two regions has been exacerbated since China further opened the coastal region in 1984. For instance, the economic growth rate of the Eastern region was 12.1 percent on average from 1984 to 1995, compared with an 9.1 percent for the Western region. This has contributed to an increased economic gap between the two regions.

In 1995, the GDP per capita of the Eastern region was 2.4 times that of the Western region. Consequently, the inter-regional economic gap has widened. This can be attributed to a number of factors such as economic reforms and open-door policy in favor of the Eastern region, different economic structure and resource conditions, coast-oriented regional policy, and the differential impact of foreign trade and foreign investment.

This chapter provides an in-depth study of the regional dualistic growth pattern of the Chinese economy and explores factors behind the uneven development between the Eastern and Western regions. It focuses on the discussion of economic structural characteristics and the impact of DFI and economic openness on regional economic growth and income inequality. It also provides a regression analysis using a time-series and cross-section model, covering 19 provinces for a 12-year period (1984-95).

7.2   Growth Pattern: Regional Dualism

Economic dualism is a typical structural feature of developing countries especially for those countries at a lower level of development. It refers to the coexistence of two disparate structural segments within the geographical territory of a country (Dutta, 1991). Dualism may take various forms such as regional dualism, urban and rural dualism, and industrial dualism. At a less-developed stage especially during the period of transition from non-market to a market economy, a certain degree of dualism may be inevitable for many countries. As the economy passes a certain stage of development, the degree of dualism may eventually diminish, resulting in the convergence of development between regions or sectors. This pattern has been the experience of many developed and newly industrialized countries.

Among recent studies on dualistic development in developing economies are Dow (1990), Dutta (1991), Patterman (1992), Zhang (1993), and Hazari and Sgro (1987). These studies provide theoretical discussion of economic dualism in the context of developing countries in general and /or provide empirical examinations of one country in particular. Some of these studies have found that for a given period, economic policies, especially regional or industrial development strategies pursued by the government of a country, can play an important part in increasing or decreasing economic dualism.

Traditionally, China's economic development has been characterized markedly by regional dualism, that is the existence of the relatively developed Eastern (coastal) region and the less-developed Western (far inland) region. During the Maoist period (1949-76), the inter-regional disparity between the two regions was to some extent reduced as a result of resource transfers from the Eastern region to the Western region. This was an important component of the Chinese government strategies for economic development and national security. In the reform or post-Maoist period, inter-regional economic disparity and income inequality have been enlarged. Regional dualism has become a basic feature of the contemporary Chinese economy.

The most important indicator demonstrating inter-regional economic disparity is the increased income gap between the two regions. During the 1978-95 period, GDP per capita in the Eastern region grew at 9.6 percent per year on average (at 1978 constant prices), compared to 7.8 percent for the Western region. As a result, the inter-regional gap of income per capita has been further widened. The trend of income inequality between the two regions has become more evident since China further opened its eastern (coastal) region to foreign investment in 1984.

Table 7.1 GDP Per Capita in the Eastern and Western Regions (*yuan*)

	GDP per capita (1978 price))			GDP per capita (current price)		
	Eastern	Western	E/W	Eastern	Western	E/W
1978	448	264	1.848	448	264	1.848
1979	534	282	1.894	541	294	1.840
1980	566	300	1.887	601	321	1.872
1981	578	317	1.823	635	344	1.846
1982	618	347	1.781	695	381	1.824
1983	672	371	1.811	766	425	1.805
1984	784	423	1.853	928	507	1.830
1985	885	477	1.855	1108	591	1.875
1986	942	501	1.880	1242	643	1.932
1987	1035	540	1.917	1447	731	1.979
1988	1143	588	1.944	1784	900	1.982
1989	1167	599	1.948	1983	970	2.044
1990	1202	621	1.936	2116	1104	1.917
1991	1298	665	1.952	2412	1231	1.959
1992	1518	731	2.077	3001	1422	2.110
1993	1789	804	2.225	4153	1785	2.327
1994	2060	876	2.352	5680	2443	2.325
1995	2315	942	2.458	7104	2948	2.410
1984-95 growth %	9.59	7.77	1.69	17.06	15.25	1.57

Source: Provincial statistical yearbooks for 19 provinces over the 1984-95 period and Statistical Yearbook of China 1995 and 1996. The provinces in Eastern Region include Guangdong, Fujian, Jiangsu, Zhejiang, Shanghai, Shandong, Hebei, Beijing, Tianjin and Liaoning. The provinces in Western Region are Shaanxi, Sichuan, Qinghai, Yunnan, Guizhou, Gansu, Ningxia, Inner Mongolia and Xinjiang.

As shown in Table 7.1, GDP per capita at 1978 constant prices increased from 784 *yuan* in 1984 to 2315 *yuan* in 1995 in the Eastern region, with an annual growth rate of 10.9 percent being reached. In comparison, the GDP per capita in the Western region during the same period increased from 423 *yuan* to 942 *yuan*, growing at 7.9 percent per year. The remarkably differential growth rates of GDP and GDP per capita resulted in a larger income gap between the two regions. In 1984, the East/West GDP per capita ratio was 1.853. By 1995, this ratio increased to 2.458, indicating that the income inequality aggravated between the Eastern and Western regions.

Population growth is another direct factor influencing income per capita in each region. As shown in Appendix Table 7.1, the annual average growth rates of the population in the Eastern and Western regions were 1.54 percent and 1.46 percent respectively during the 1984-95 period. Since the population grew slightly faster in the Eastern region than in the Western region, it did not contribute to the widening income per capita gap between the two regions.

Consequently, it is the regional divergence in economic growth rates that can be identified as the principal cause for the inter-regional income disparity. In fact, many factors contributed to the differential growth rates of the economy in the two regions. Apart from the government regional development strategy and policies that were favorable to the Eastern region, structural factors and differential economic openness, especially the coast-oriented distribution of direct foreign investment, has played a crucial role in the rise of the dualistic growth pattern of the Chinese economy.

7.3 Structural Characteristics

The dualistic growth pattern of the Eastern and Western regions is to a large extent attributed to the various structural characteristics of the two regions. Three such characteristics are industrial structure, openness of the economy, and investment structure. The different structural characteristics in the two regions shape the growth conditions and disparate growth pattern, resulting in income inequality between the two regions. The income gap in turn further reinforces the disparity of capital formation and regional dualistic growth pattern.

7.3.1 Industrial Structure

The industries in an economy can be broadly classified into three categories, viz., primary, secondary and tertiary (service) industries. In general, primary industry constitutes a large share of GDP and total employment when the economy is at a lower stage of development. As economic development in a country or region reaches a higher level, the secondary and tertiary industries

120 *Foreign Investment and Economic Development in China*

tend to become more important and account for large shares in GDP and total employment.

In the case of China, the industrial structure has experienced remarkable changes since the later 1970s. The share of primary industry in GDP declined from 28.4 percent in 1978 to 20.9 percent in 1995 while the share of secondary industry rose from 48.6 percent to 49.2 percent during the same period. The tertiary industry's share of GDP increased from 23 percent to 31.6 percent. Although structural change occurred in all regions, the pace and extent to which the economic structure changed are significantly different between the Eastern and Western regions.

Table 7.2 Industrial Structure of the Eastern and Western Regions

	Eastern Region			Western Region		
	Primary Industry as % of GDP	Second Industry as % of GDP	Tertiary Industry as % of GDP	Primary Industry as % of GDP	Second Industry as % of GDP	Tertiary Industry as % of GDP
1984	27.2	50.7	22.1	37.3	39.0	23.7
1985	25.1	52.5	22.4	35.6	40.7	23.7
1986	24.1	47.6	24.4	34.5	40.6	24.9
1987	23.2	50.8	24.9	34.4	39.8	25.7
1988	23.3	51.6	25.6	34.1	40.8	25.1
1989	22.6	51.3	29.0	33.9	41.3	26.3
1990	23.2	49.5	7.2	35.7	38.2	26.0
1991	21.8	49.8	28.3	34.2	39.4	26.4
1992	18.8	52.0	28.7	31.4	41.4	27.2
1993	15.6	51.1	33.3	26.0	41.9	32.1
1994	15.8	50.6	33.5	27.2	41.4	31.3
1995	16.0	49.8	34.2	26.8	41.5	31.7

Source: Provincial statistical yearbooks for 19 provinces over the period 1984-95, and Statistical Yearbook of China 1995-97.

As shown in Table 7.2, the share of the primary industry in the GDP of the Eastern region declined from 27.2 percent in 1984 to 16 percent in 1995, in comparison with a corresponding decrease from 37.3 percent to 26.8 percent in the Western region. Although the secondary and tertiary industries increased their importance in the Western region, their shares in GDP are still lower than that of the Eastern region. By 1995, the secondary industry produced 41.5 percent of GDP in the Western region compared to 49.8 percent in the Eastern region. Similarly, the tertiary industry contributed 31.7

percent of GDP in the Western region compared to 34.2 percent in the Eastern region.

The industrial structure tends to significantly affect the overall economic growth rate of each region reflecting the different industrial makeup. For example, during the 1984-1995 period the GDP produced by the tertiary industry grew at 16.6 percent annually in the Eastern region compared to 12.1 percent in the Western region. The GDP produced by the secondary industry grew at 11.8 percent in the Eastern region compared to 9.7 percent in the Western region. The growth rates of the primary industry were similar in the two regions (6.0 percent and 5.7 percent respectively). As the secondary and tertiary industries produced over 84 percent of GDP in the Eastern region, their relatively higher growth rates contributed to a higher rate of GDP growth. In the Western region, however, these two industries produced only 73 percent of GDP.

7.3.2 Openness of the Economy

The openness of an economy can be measured by the sum of exports and imports as a percentage of GDP in an economy. Openness increasingly being viewed as an important attribute to ensure steady economic growth in less developed countries, especially following the successful experiences of East and Southeast Asian regions.[2] In the case of China, the openness of the economy has increased sharply since the Chinese Government started pursuing economic reforms including the 'opening up' of the economy in 1979. The sum of exports and imports as a percentage of GDP has risen from 10.3 percent in 1979 to 22 percent in 1996 (calculated at 1984 constant prices and the 1984 constant exchange rate). Several studies (e.g. Kwan and Cotsomitis, 1985; Fan, 1992; Lardy, 1994; and Sun, 1995, 1996) have proven that export expansion has become an important driving force for the rapid economic growth of China in the past two decades.

However, as the Government's 'open' policy has been primarily oriented to the Eastern region, the openness of the Chinese economy is subject to a remarkably regional disparity. As shown in Table 7.3, during the period from 1984 to 1996, both the Eastern and Western regions made a considerable progress in opening their economies. However, there is still considerable disparity in the economic openness between the two regions. For instance, 17 percent of the total output of the Eastern region was sold in overseas markets in 1995, in contrast to 3.1 percent of output from the Western region.

Greater openness especially the expansion of exports can be expected to promote economic growth in various ways. First, exports guide countries (or regions) to produce products in which they have comparative advantage. By specializing in such industries or products, a country or region can achieve higher productivity and a lower real cost of production.

Table 7.3 Openness of the Economy in the Eastern and Western Regions

Year	Eastern Region				Western Region			
	Exports ($ mill.)	Exports as % of GDP	Imports ($ mill.)	(X + M) as % GDP[1]	Exports ($ mill.)	Exports as % of DP	Imports ($ mill.)	(X + M) as % of GDP
1984	18643	12.3	3683	14.7	793	1.4	479	2.3
1985	20081	11.2	7327	15.3	1065	1.6	608	2.6
1986	19399	9.8	7568	13.6	1445	2.0	794	3.1
1987	24348	10.7	9362	14.8	2024	2.5	968	3.7
1988	28402	10.7	13457	15.8	2554	2.6	1283	4.1
1989	31034	10.9	13822	15.8	2810	2.8	1457	4.3
1990	36997	11.8	13386	16.0	3156	2.9	1147	3.9
1991	42908	11.9	17825	16.8	3713	3.0	1165	3.9
1992	53475	12.5	24690	17.8	4520	3.2	2125	4.7
1993	68393	12.7	40001	20.1	5302	3.3	3545	5.4
1994	103604	16.1	98857	31.5	4933	2.7	6026	6.0
1995	128102	17.0	113980	32.2	6322	3.1	6786	6.4
1996	131990	16.0	123430	30.0	5610	2.4	6360	5.2

Notes: 1. The sum of exports and imports as share of GDP = the sum of exports and imports converted into the Chinese currency (*Renminbi*) / GDP. In this calculation, the exchange rate is the 1984 official exchange rates (US$1 = 2.327 *Renminbi*), the values of GDP are in the 1984 constant price, and the values of export and import are in the 1984 US constant price. The figures of exports and imports in this Table are in current dollar.

Source: The data for the period from 1984 to 1993 are from the Statistical Yearbooks of 10 Eastern provinces and 9 Western provinces for years 1985-94. The data for 1994-96 are from State Statistical Bureau (SSB) of China, Statistical Yearbook of China 1994, 1995, 1996 and 1997.

Trade can also induce economic resources to flow from less productive sectors to sectors with comparative advantage, thus increasing the overall allocative efficiency of resources. In addition, trade facilitates a country or region to access new technology and overseas markets, which in turn can be expected to have positive effects on productive and management efficiencies.

A greater openness and integration into the world economy is one of principal factors propelling the economic growth of China's eastern region in the past two decades.

Due to the much smaller share of foreign trade making up its GDP, the impact of exports on economic growth in the Western region is less significant than in the Eastern region. Although exports grew slightly faster in the Western region than in the Eastern region, the exports per capita in the Eastern region were far larger. This figure was US$44.8 in the Eastern region in 1980, compared to US$1.2 in the Western region. In 1995, it reached US$284.3 in the Eastern region compared to only US$21.1 in the Western region (SSB, 1981-96). The different degrees of openness of the economy in the two regions are an important contributing factor in the differential economic growth rates and thereby the resulting income gap between the regions.

7.4 Foreign Investment and Capital Formation

Since the early 1980s, direct foreign investment has played increasingly important role in China's capital formation and economic growth. From 1984 to 1996, the arrived DFI in China grew at an annual average 33.9 percent, with the accumulative total value increasing from US$1,258 million to US$41,726 million. In 1995, foreign capital accounted for 11.5 percent of the total fixed capital investment of China (SSB, 1996). DFI has thus become a dynamic part of domestic capital formation, promoting Chinese economic growth. However, the regional distribution of foreign investment is highly uneven. During the period from 1979 to 1995, over 90 percent of DFI was in the Eastern (coastal) region, with only 6 percent in the Central region and 4 percent in the Western region. Therefore, one can easily see the differential impact of DFI on economic growth in different regions.

The regional distribution of DFI is primarily determined by the investment environment and return on capital in different regions. The investment environment in a region is affected by a number of social and economic factors, which in the theory of international investment are termed *location-specific factors*. These include infrastructure, transportation, economic structure and the development level, economic policy, legal system and resource endowment. In the case of China, *location-specific factors* differ considerably between the Eastern and Western regions. In general, the

Eastern region is more economically developed, with considerably superior infrastructure, especially in its transportation and communication systems.

Traditionally, China's major economic centers including major seaports and airports have been concentrated in the eastern coastal region. Better service facilities and human resources make the investment environment in the Eastern region superior to that of the Western region. In terms of economic structure, the Eastern region is also more developed than the Western region, with the manufacturing and tertiary industries having led the economic growth in this region.

The existing advantages of the Eastern region over the Western region have been reinforced by coast-oriented economic reforms and open door policy since the early 1980s. The Chinese government committed a large amount of capital to the Eastern region to improve infrastructure, such as transportation, communications, public utilities and service facilities. Furthermore, the government granted a package of preferential policies to the Eastern region, including favorable policies on taxation, foreign trade and investment, and more autonomy in economic decision making, all of which resulted in a favorable investment environment.

In particular, the open-door policy has been pursued with a remarkable spatial dimension. In 1979, the Chinese government initiated the open-door policy by establishing four Special Economic Zones (SEZs) in the southeast coastal region. These SEZs were initially designed as laboratories for the use of foreign investment, where special economic policies were adopted. The spatial proximity to Hong Kong, the source of about 80 percent of total DFI in China, was important. In the initial stage, DFI was highly concentrated in the four SEZs (Shenzhen, Zhuhai, Shantou and Xiamen) and during the period from 1979 to 1983, around 80 percent of total DFI projects was located in the four SEZs. However, as the government used administrative regulations to isolate the SEZs economically from the rest of the country, the SEZs in essence became foreign enclaves with few economic linkages with other regions.

In 1984, 14 coastal cities were opened to foreign investment. As in the SEZs, a series of special economic policies were introduced in these open coastal cities (OCCs). This helped DFI to diffuse spatially from the SEZs to the fourteen OCCs across ten coastal provinces. Consequently, the contracted DFI in the fourteen OCCs exceeded that in the SEZs. Since 1986 DFI has gradually spread to the other regions including the other coastal areas and the vast inland regions. In 1990, the new emphasis of open policy shifted to the Shanghai Pudong New Area, Changjiang (Yangtse River) Delta and Minnan Delta. This was followed by a rapid expansion of DFI to the inland regions after Deng Xiaoping's 'south tour' in early 1992. As a result, DFI diffused quickly to the inland regions and spread widely across the country.

However, due to the favorable investment environment in the Eastern region, DFI has been concentrated in this region since the beginning of open-door policy. As Table 7.4 indicates, during the period of 1983-1996, 87.9 percent of the arrived DFI in China was located in the Eastern region, with 8.8 percent and 3.4 percent in the Central inland and the Western inland regions respectively. Although DFI has gradually spread to the inland regions since the mid-1980s, the Eastern region is still the primary location of foreign investment.

Table 7.4 Regional Distribution of Direct Foreign Investment(US$ millions)

	1983-85[1]		1986-89		1990-1996		1983-96	
Region	Value	%	Value	%	Value	%	Value	%
All Regions	2780	100.0	8498	100.0	154878	100.0	166155	100.0
Eastern	2514	90.4	7510	88.4	135976	87.8	146024	87.9
Guangdong	1701	61.2	3440	40.5	45829	29.6	50871	30.6
Fujian	186	6.7	572	6.7	16890	10.9	17647	10.6
Jiangsu	63	2.3	261	3.1	18807	12.1	19131	11.5
Zhejiang	37	1.3	123	1.5	5341	3.4	5501	3.3
Shanghai	161	5.8	1017	12.0	13280	8.6	14457	8.7
Shandong	43	1.5	218	2.6	11008	7.1	11269	6.8
Beijing	200	7.2	1057	12.4	5543	3.6	6799	4.1
Tianjin	71	2.5	217	2.5	5358	3.5	5645	3.4
Hebei	14	0.5	58	0.7	2490	1.6	2526	1.5
Liaoning	39	1.4	340	4.0	6974	4.5	7353	4.4
Hainan	41	1.5	251	3.0	4208	2.7	4500	2.7
Central	146	5.3	532	6.3	13881	9.0	14559	8.8
Western	96	3.4	456	5.4	5021	3.2	5572	3.4

Notes: 1. The figures for 1983-85 includes direct foreign investment and other forms of investment such as compensation trade, processing assembly and international leasing. The figure since 1986 refer only direct foreign investment.

Source: *Almanac of China's Foreign Economic Relations and Trade, 1984-1994; Statistical Yearbook of China 1991-1997*; and China Foreign Economic Statistics 1979-1992, 1994 and 1996.

For example, Guangdong Province is the single largest recipient of DFI. Over the period from 1983 to 1996, DFI in this province amounted to US$50.9 billion, which accounted for 30.6 percent of the national total. In the initial stage (1983-85), 61 percent of total DFI flowed into Guangdong. The province has remained the most important location for DFI although its share in the national total has declined since the mid-1980s. Jiangsu Province is second only to Guangdong, with DFI amounting to US$19.1 billion during the 1983-96 period and accounting for 11.5 percent of total DFI. This is followed by Fujian Province whose DFI amount accounted for 10.6 percent of the national total.

Table 7.5   Direct Foreign Investment as a Share of Total Capital Formation [1] (%)

	1984	1986	1988	1990	1991	1992	1993	1994	1995
Eastern Region:	1.9	2.6	3.6	6.1	7.5	11.2	17.6	24.8	22.3
Guangdong	9.7	10.3	9.7	18.7	20.3	21.2	26.4	37.8	38.2
Fujian	3.3	3.5	5.2	12.7	25.9	39.6	44.5	59.5	51.0
Jiangsu	0.4	0.3	1.0	1.9	2.8	10.9	14.7	24.4	25.4
Zhejiang	0.3	0.5	0.6	1.2	2.0	4.5	7.6	9.8	7.3
Shanghai	1.1	3.5	5.5	3.7	3.6	12.2	28.3	19.6	15.6
Shandong	0.1	0.3	0.4	2.1	2.2	0.7	11.9	19.9	17.7
Hebei	0.1	0.2	0.3	1.1	1.6	2.9	5.0	6.2	5.2
Beijing	0.4	5.1	13.4	7.4	9.0	9.6	9.2	23.3	10.8
Tianjin	1.0	2.1	1.1	4.5	3.9	7.5	13.8	27.6	33.1
Liaoning	0.1	0.7	1.3	4.5	5.2	5.5	10.2	14.3	14.2
Western Region	0.2	0.6	0.9	0.6	0.4	0.9	2.5	5.7	3.9

Notes: 1. Direct foreign investment was initially in US dollar in Chinese statistics, and is here converted into Chinese *yuan* when calculating the share of DFI in the total capital formation. The exchange rates are official rates in each year. The capital formation refers to total fixed capital investment in each year.

Source: Statistical Yearbooks of the 19 provinces (10 Eastern provinces and 9 Western provinces) for the period from 1985 to 1995, and *Statistical Yearbook of China 1994-1997*.

However, both the Central and Western regions received a very small share of DFI. These two inland regions (including 19 provinces) received less than 40 percent of the DFI which had flowed into Guangdong, the single province, during the 1983-96 period. DFI which flowed into the Western region (9 provinces) was amounted to only 11 percent of DFI in Guangdong Province over the same period. Clearly, the regional distribution of DFI in China has been highly uneven.

Table 7.5 indicates the differing contribution of DFI to total capital formation in the two regions. In some eastern provinces especially those in the Southeast region, by 1995 DFI had overtaken the government investment budget and become the largest capital source of total fixed capital investment. For instance, DFI as a share of the total fixed capital investment in Fujian and Guangdong was 51 percent and 38.2 percent respectively in 1995.

In other coastal provinces, the contribution of DFI to total capital formation has also been important. In Tianjin and Jiangsu, DFI as a share of total fixed capital investment was 33.1 percent and 25.4 percent respectively. As a result of the large foreign capital inflows, the investment rate (total fixed capital investment as a share of GDP) in the Eastern region has been constantly higher than in the Western region. This is shown in Table 7.6.

Table 7.6 Fixed Capital Investment and its Share of GDP

	Eastern Region		Western Region	
	Investment[1]	As % of GDP	Investment	As % of GDP
1984	872	0.25	311	0.24
1985	1247	0.29	469	0.28
1986	1286	0.31	550	0.29
1987	1855	0.32	670	0.28
1988	2316	0.32	647	0.28
1989	2115	0.26	617	0.24
1990	2294	0.26	720	0.23
1991	3724	0.27	909	0.26
1992	4294	0.33	1240	0.31
1993	7250	0.41	1809	0.35
1994	9813	0.39	2177	0.30
1995	12369	0.39	2643	0.30
1984-95	48635	0.35	12587	0.29

Note: 1. The unit for investment is 100 million *yuan*. Both GDP and fixed capital investment are in current price in each year.

Source: The provincial statistical yearbooks of the 10 eastern provinces and 9 western provinces for the 1985-96 period.

As a result of its uneven regional distribution, the contribution of DFI to total capital formation is significantly different between the Eastern and Western regions. Foreign capital as a share of the total capital reached 22.3 percent in the Eastern region in 1995, which is much higher than the correspondent percentage in the Western region (3.9 percent). Table 7.5 shows the changing role of foreign investment in the total capital formation in the two regions during the 1984-95 period.

During the 1984-95 period, fixed capital investment as a share of GDP was 34.6 percent on average in the Eastern region, while the investment rate in the Western region was just 29.1 percent on average. As shown in Table 7.6, there is no single year over this period in which the investment rate of the Western region was higher than that of the Eastern region. The Eastern region has continually headed the Western region in capital formation and this has been the most important determinant of the differential rates of economic growth in the two regions.

Apart from the impact of foreign investment, domestic capital flows from the Western region to the Eastern region is also a factor influencing regional capital formation. Due to preferential economic policies pursued by the Chinese government and the favorable investment environment in the Eastern region, companies and government agencies in the Western region tend to make investment in eastern provinces, especially open coastal cities or provinces such as the five Special Economic Zones (Shenzhen, Zhuhai, Xiamen, Shantou and Hainan), Guangdong, Fujian and Shanghai, with view to achieving higher returns on capital. This results in domestic capital flows from the Western region to the Eastern region.

For example, Shaanxi Province invested over 270 million *yuan* in eastern provinces in 1992, resulting in a net capital outflow of 225 million *yuan* (Shaanxi Statistical Bureau, 1994). In the past few years, some other Western provinces have also experienced net capital outflows. Such inter-regional flows of domestic capital have a differential impact on capital formation in the two regions. They increase capital supply and stimulate capital formation in the Eastern region. However, capital outflows negatively affects capital formation in the Western region. This contributes to different investment rates in the two regions.

In short, foreign investment, through its contribution to domestic capital investment and its positive effects on technology transfer and export growth, has accelerated the economic growth of the Eastern region. By contrast, the contribution of foreign investment to capital formation in the Western region has been slight, and hence its impact on the economic growth of the region is less significant.

7.5 Rural Industry Development and Inter-regional Economic Disparity

Rural industry growth has been an important factor promoting the economic development of China in the past two decades. It has become the most dynamic part of the Chinese economy and plays an increasing role in economic growth, especially in rural development. As a principal indicator of rural industry, industrial enterprises run by townships and villages developed dramatically in the past 17 years. During the period 1978 to 1995, the total output value of township and village enterprises (TVEs) grew at annual average 24.8 percent (at 1978 constant prices). This growth rate is almost 10 percentage points higher than the growth rate of the gross output value of China's industry.

However, the development of rural industry is also subject to regional imbalance. As Table 7.7 shows, the total output value of TVEs in the Eastern region increased from 120.4 billion *yuan* in 1985 to 1221.3 billion *yuan* (at 1984 constant prices) in 1995, with an annual growth of 26.1 percent. During the same period, the total output value of TVEs in the Western region increased from 16.1 billion *yuan* to 133.9 billion *yuan* (at the 1984 constant price), with an annual growth of 23.5 percent. In terms of labor productivity, the value added produced by each worker in industrial TVEs was 16,046 *yuan* in the Eastern region, some 60.5 percent higher than that of the Western Region (9,997 *yuan*).

The differential growth of rural industry aggravates inter-regional economic disparity, especially the income per capita gap. This is because rural industry provides the primary source of income for the rural population, which makeup 80 percent of the total population of China. As shown in Table 7.7, due to the different growth of rural industry, the inter-regional gap of TVEs' output value per capita in 1995 has been further widened from 250.7 *yuan* in 1985 to 2263.6 *yuan* (at 1984 constant price). In terms of value added, the TVEs' value added per capita was 1526 *yuan* in the Eastern region in 1995, compared to 322 *yuan* in the Western region.

Rural industrial growth has contributed significantly to rural economic development and income growth in the Eastern region where industrial enterprises run by townships and villages have now become dominant players in rural economic growth. Since 1987, the industrial output value produced by TVEs in this region has surpassed the gross output value of the agriculture sector (including farming, forestry, animal husbandry and fishing). By 1995, it was 3.86 times as much as the gross output value of the agriculture sector. The rapid growth of rural industry has played a leading role in rural economic development of the Eastern region. In the Western region, however, the rural industry has developed relatively slower and its role in promoting the overall economic growth has been less important. By 1995, the industrial output value produced by TVEs was still less than that of the agriculture sector in the region.

Table 7.7   Total Output Value of Township and Village Enterprises[1]

(Unit: billion *yuan*)

	Eastern Region			Western Region		
	Output value I[2]	Output value II	Output value per capita[3]	Output value I	Output value II	Output value per capita
1985	127.6	120.4	313.1	16.6	16.1	62.4
1986	157.2	140.4	360.6	19.4	18.2	69.7
1987	210.2	178.1	450.3	24.6	21.8	82.2
1988	310.9	235.5	587.2	33.3	26.0	96.5
1989	352.7	246.7	607.4	38.6	28.6	104.6
1990	383.0	258.8	623.1	41.9	28.3	101.7
1991	484.2	310.4	732.2	51.9	33.7	119.8
1992	750.4	451.9	1054.5	77.3	47.9	168.9
1993	1278.6	672.9	1550.6	140.6	76.0	262.2
1994	2389.2	999.2	2248.4	211.0	88.2	300.6
1995	3568.4	1221.3	2710.6	391.4	133.9	447.0
1985-95 annual growth %	39.5	26.1	24.1	37.2	23.5	21.8

Notes: 1. The figures in this Table refer to industrial enterprises run by townships and villages in the rural areas. 2. The output value I refers to the output value at current prices, while the output value II refers to the output value calculated at the 1984 constant price. 3. The output value per capita = output value at the 1984 constant price / population. The unit for the output value per capita is *yuan*.

Source: SSB. *The Statistical Yearbook* of *China's Agriculture* 1986-1996.

In terms of the income effect, the differential development of the rural industry has contributed greatly to the income gap between the Eastern and Western regions, especially in rural areas. For example, in 1995 there were

42.85 million rural workers employed by industrial TVEs in the Eastern regions, which accounted for 26.3 percent of the total rural labor force. The total number of employees in all types of TVEs reached 63.72 million people, accounting for 39.1 percent of the total rural labor force. A massive participation of rural labor force in TVEs' industrial and commercial activities contributed directly to the income growth of rural population in the Eastern region. For instance, the average wage income earned by each worker in TVEs in the region was 2064 *yuan* in 1995. On average each worker in TVEs produced a pre-tax operating profit of 2933 *yuan*.

By comparison, the contribution of TVEs to the economic development and income growth of the Western region has been less significant. There were 9.66 million people working in industrial TVEs in 1995, accounting for only 8.4 percent of the total rural labor force. With the inclusion of employees in TVEs of other sectors, the total employees in the TVEs of the Western region numbered 21.17 million people, which accounted for 18.4 percent of the total rural labor force.

However, in terms of income effect, the annual average wage for employees in TVEs was 831 *yuan* in 1995, which was only 40 percent of that of the Eastern region. Likewise, the pre-tax operating profit of TVEs in the Western region was 1101 *yuan* in 1995, which is also much less than the corresponding indicator for the Eastern region. This indicates that the differential development of rural industry in the two regions has further contributed to the inter-regional income inequality.

7.6 Regression Analysis

In the previous section, a discussion of the determinants of inter-regional economic disparity from a structural perspective has been provided. It has highlighted the contributions that different industrial structure and openness of the economy, coast-oriented DFI, divergent conditions for capital formation, and the different levels of rural industrial development made to enlarging inter-regional economic disparity. This section investigates fundamental factors determining the divergent economic growth of the Eastern and Western regions, within a production function framework using econometric modeling techniques.

7.6.1 Model

The production function to be measured in this model is assumed to have the properties of the Cobb-Douglas production function, a basic and widely used form of production function. Due to the nature of data, all variables are aggregates at the provincial level. Output is expressed by gross domestic product (GDP), and inputs include capital and labor. Capital input can be divided into two components: domestic and foreign capital. In this model, domestic capital input (DK) is expressed by domestically financed fixed investment. Direct foreign investment (DFI) represents foreign capital. It can also be regarded as an important indicator of the openness of the economy. In addition, as DFI is highly related to the transfer of foreign technology and management skills, thereby promoting technological progress in joint ventures and other firms, it can capture the impact of technology on production to some extent.[3] Therefore, the production function can be written as:

$$GDP = A \, DK^{\alpha} \, DFI^{\beta} \, L^{\gamma}$$

In logarithmical form, the model can be written as below:

$$ln \, GDP = c + \alpha \, ln \, DK + \beta \, ln \, DFI + \gamma \, ln \, L + u$$

where GDP = gross domestic product,
    c = log (A),
    DK = domestically financed fixed investment,
    DFI = direct foreign investment,
    L   = the number of labor employed in production, and
    u = stochastic error term.

In terms of econometric methodology, the model used here is a Kmenta Model (Kmenta 1986), that is a cross-section and time-series model. This model is suitable and efficient for a regression analysis using cross-section and time-series data since it takes a cross-sectional heteroskedasticity and timewise autocorrelation into account, producing a reliable econometric output. To measure the variation in the effect of independent variables on the dependent variable (GDP) between the Eastern and Western regions, a dummy variable (D) is used. It takes the value 1 for the Eastern region and 0 for the Western region. Thus the regression formula is expressed as follows:

$$ln\ GDP = c_1 + c_2\ D + \alpha_1\ ln\ DK + \alpha_2\ ln\ DK\ (D) + \beta_1\ ln\ DFI$$
$$+ \beta_2\ ln\ DFI(D) + \gamma_1\ ln\ L + \gamma_2\ ln\ L\ (D)$$

where c_1 is the intercept for the Western region; c_2 is the differential intercept for the Eastern region. The coefficients α_1, β_1 and γ_1 are the estimated elasticity of GDP with respect to DK, DFI and L in the Western region. α_2, β_2 and γ_2 are the differential coefficients of DK, DFI and L for the Eastern region. If α_2, β_2 and γ_2 are significantly different from zero, the effects of changes in DK, DFI, and L on GDP growth will be significantly different between the Eastern and Western regions, i.e. the responsiveness of GDP to changes in DK, DFI and L will be significantly different between the two regions. The regression can also be run separately for each region without a dummy variable; the results are the same as those obtained by using a regional dummy variable.

It is hypothesized that increases in domestic investment, direct foreign investment and labor will affect GDP growth positively, and therefore, the coefficients of DK, DFI and L are expected to be positive in both the regions. However, the impact of DFI on GDP growth is expected to be significantly different between the two regions, being stronger in the Eastern region than in the Western region. This is because DFI has been concentrated in the Eastern region during the past 18 years

7.6.2 The Data and Spatial Spread

The data used in this regression analysis are the provincial data of 19 provinces for the 1984-95 period. The spatial coverage is of 10 provinces in the Eastern region (Guangdong, Fujian, Jiangsu, Zhejiang, Shanghai, Shandong, Hebei, Beijing, Tianjin and Liaoning) and 9 provinces in the Western region (Shaanxi, Sichuan, Qinghai, Yunnan, Guizhou, Gansu, Ningxia, Inner Mongolia and Xinjiang). The sources of data for the period 1984-93 are from provincial statistical yearbooks for each province, with data for 1994 and 1995 being from *the Statistical Yearbook of China* 1995 and 1996. As the data include 19 provinces over a 12-year period, each variable has 228 observations.

In order to remove the influence of inflation on the variables and their relations, *GDP*, *DK* and *DFI* are expressed in constant prices (1984=100). The current values of *GDP* and *DK* are converted into real values using the 1984 constant prices for each province. The values of DFI, which are originally expressed in current U.S. dollar, are deflated using the U.S. GDP implicit price deflators published in *Survey of Current Business* (US Department of Commerce, 1997).

7.6.3 Regression Results and Implications

The two sets of regression results are presented in Table 7.8.

Table 7.8 Production Function Estimates for the Eastern and Western Regions

Variables Name	Estimated Coefficient	T-Ratio	Partial Correlation	Standardized Coefficient
ln DK	0.5230	17.34	0.760	0.5066
ln DK (D)	-0.0487	-1.081*	-0.087	-0.1285
ln DFI	0.0356	4.192	0.272	0.0884
ln DFI (D)	0.0558	3.957	0.258	0.1598
ln L	0.4806	19.46	0.795	0.4396
ln L (D)	-0.1037	-2.878	-0.190	-0.1549
Constant	1.8717	24.47	0.855	0.0000
Constant (D)	0.9424	7.504	0.451	0.4668

Buse $R^2 = 0.9801$, d.o.f = 220; F-statistics = $1550.83 > F_{0.01} = 2.41$ (for one tailed test).

The Eastern Region:

ln DK	0.4743	11.52	0.736	0.4513
ln DFI	0.0914	8.134	0.603	0.2620
ln L	0.3769	14.39	0.801	0.4440
Constant	2.8141	28.30	0.935	0.0000

Buse $R^2 = 0.9494$, d.o.f. = 116, F-statistics = $725.27 > F_{0.01} = 3.48$ (for one tailed test).

The Western Region:

ln DK	0.5230	17.30	0.862	0.4551
ln DFI	0.0356	4.182	0.380	0.0741
ln L	0.4806	19.43	0.885	0.5225
Constant	1.8717	24.42	0.923	0.0000

Buse $R^2 = 0.9777$, d.o.f = 104; F-statistics = $1522.78 > F_{0.01} = 3.48$ (for one tailed test).

* With exception of the coefficient for *ln DK* (D), the t-ratios for all coefficients are statistically significant at 1% level (for one tailed test).

The results indicate:

1. The effect of DFI on GDP growth is positive in both the Eastern and Western regions. However, there is a significant difference in the impact of DFI between the two regions, being stronger in the Eastern region than in the Western region. This supports the hypothesis about the effect of DFI on regional economic growth. In terms of the elasticity coefficient, a one percent increase in DFI would generate a 0.0914 percent growth of GDP in the Eastern region, while a one percent change in DFI would lead to a 0.0356 percent change in the GDP of the Western region. As the elasticity coefficient reflects the percentage change in GDP in response to a one percent change in DFI, its low level implies that the percentage change in DFI is much larger than that of GDP. This is true. During the period from 1984 to 1995, GDP grew at 12.1 percent per annum in the Eastern region, while DFI grew at 37.1 percent per annum. In the Western region GDP and DFI grew at 9.1 percent and 34.9 percent respectively. This suggests that to estimate the actual impact of DFI on GDP growth, the actual growth rate of DFI must also be considered.

2. Domestically financed investment is the most important determinant of economic growth in both the Eastern and Western regions. The elasticity coefficients of DK are statistically significant, indicating that the responsiveness of GDP to changes in domestically financed fixed investment is positive and strong in the two regions. A 1 percent growth in DK will lead to a 0.394 percent growth of GDP in the Eastern region, and to a 0.523 percent growth of GDP in the Western region. A higher elasticity coefficient of DK for the Western region suggests that the economic growth in the region is more responsive to changes in domestically financed investment. This indicates that the economic growth of the Western region is more dependent on domestic investment in comparison to the Eastern region.

3. There is a significant difference in the elasticity of GDP with respect to labor change between the two regions. A 1 percent increase in labor force is associated with 0.481 percent growth of GDP in the Western region, while the corresponding coefficient for the Eastern region is 0.377 percent. This implies that the economic growth of the Western region is more dependent on labor input, and the production process is more labor-intensive compared to that of the Eastern region.

4.   The differential intercept of production function for the Eastern region
     is significantly different from zero (0.9432), indicating that other factors
     such as technology, economic structure and management have played a
     greater part in the economic growth of the Eastern region than in the
     Western region. A higher intercept in the production function (i.e. a
     higher position of the isoquant curve) suggests that technological and
     structural factors contributed more to economic growth in the Eastern
     region than in the Western region.

Using the estimated elasticity coefficients of GDP with regard to *DK*,
DFI and *L*, the relative contributions of these inputs to economic growth in
the two regions can be identified.[4] During the period of 1984-95, the average
annual growth rate of GDP was 12.05 percent in the Eastern region, and the
average growth rates of *DK, DFI* and *L* are 13.79 percent, 37.06 percent and
2.37 percent respectively. Over the same period, the GDP of the Western
Region grew at 9.12 percent on the average. *DK, DFI* and *L* grew at annual
rates of 10.80 percent, 34.9 percent and 2.40 percent respectively (SSB,
1986-96). Thus, the estimated GDP growth functions for the Eastern and
Western regions can be expressed respectively as:

(1) For the Eastern region:

$$GDP^* = 2.814 + 0.394 \, DK * + 0.091 \, DFI* + 0.474 \, L*$$
$$= 2.814 + (0.394 \times 13.79) + (0.091 \times 37.06) + (0.377 \times 2.37)$$
$$= 2.814 + 5.433 + 3.634 + 0.893$$
$$= 12.774$$

where asterisk denotes average proportional change with respect to time.

$$GDP^r / GDP^* = 12.05 / 12.774 = 0.943$$

where GDP^r is the real GDP growth rate and GDP^* is the estimated growth
rate. Therefore, the relative contributions of the factors of production to
GDP growth for the Eastern region are respectively:

$$DK^* / GDP^* = 5.433 / 12.774 = 0.425, \text{ or } 42.5\%$$
$$DFI^* / GDP^* = 3.634 / 12.774 = 0.285 \text{ or } 28.5\%$$
$$L^* / GDP^* = 0.893 / 12.774 = 0.07, \text{ or } 7.0\%$$

The contribution by other factors to GDP growth = 2.814 / 12.774 = 0.223, or 22.3%.

(2) For the Western region:

$$GDP^* = 1.872 + 0.0.523 \; DK + 0.0356 \; DFI + 0.481 \; L$$
$$= 1.872 + (0.523 \times 10.80) + (0.0356 \times 34.9) + (0.481 \times 2.40)$$
$$= 1.872 + 5.648 + 1.242 + 1.152$$
$$= 9.914$$

$$GDP^r / GDP^* = 9.124 / 9.914 = 0.9203$$

Thus, the relative contributions of the three inputs to the output are:

$$DK^* / GDP^* = 5.648 / 9.914 = 0.570, \text{ or } 57.0\%$$
$$DFI^* / GDP^* = 1.242 / 9.914 = 0.125 \text{ or } 12.5\%$$
$$L^* / GDP^* = 1.152 / 9.914 = 0.116, \text{ or } 11.6\%$$

The contribution of other factors to GDP growth in the Western region = 1.872 / 9.914 = 0.189, or 18.9%.

These results suggest that:

1. Domestically financed investment (DK) is still the largest contributor to the economic growth in both regions. In the Eastern region, 42.5 percent of GDP growth was accounted for by the increase in domestically financed investment during the period from 1984 to 1995. In the Western region, domestically financed investment is even more important in determining economic growth, contributing to 57 percent of GDP growth over the same period. This implies that the economic growth in the Western region is more dependent on domestic investment than in the Eastern Region.

2. DFI is an important factor contributing to divergent economic growth rates between the Eastern and Western regions. Over the 1984-95 period, DFI contributed 3.634 percentage points of the GDP growth rate of the Eastern region, which accounted for 28.5 percent of the economic growth (12.05 percent). In other words, nearly one-third of the GDP growth in the region was brought by DFI. DFI has become the second largest contributor to the economic growth of the Eastern region. In comparison, DFI contributed 1.24 percentage point of the GDP growth rate in the Western region, accounting for 12.5 percent of the

GDP growth (9.12 percent). This suggests that DFI alone resulted in a 2.392 percentage point deference (3.634-1.242=2.392) in the growth rates of GDP between the two regions. This can explain 81.6 percent of the overall difference in growth rates (12.05 -9.12 = 2.93) between the Eastern and Western regions (i.e. 2.392 / 2.93 = 0.816). This indicates that *DFI is the most important factor contributing to the differential economic growth and income inequality existing between the Eastern and Western regions during the reform era.*

3.   Labor force (L) contributes to economic growth differently in each of the regions. In the Eastern region, 7 percent of GDP growth was generated by the increase of labor force compared to 11.6 percent in the Western region. This suggests that economic growth in the Western region is more responsive to and more dependent on an increase in the labor force. The relative contribution of the labor force to economic growth is more significant in the less-developed Western region than in the Eastern region This also implies that labor-intensive industries account for a larger share of the economy of the Western region than that in the Eastern region.

4.   The estimated coefficient for the constant '*c*' of the production function for the Eastern region is significantly higher than that for the Western region. This points to other factors especially technology, management and economic structure having played a more important role in the economic development of the Eastern region than in the Western region.

The above findings spell out that the differential growth rates of domestically financed investment, DFI and improvement of productivity are the primary reasons for the divergent economic growth of the Eastern and Western regions. A higher investment rate, increased openness of the economy especially DFI, technological progress, macro and micro-economic reforms and industrial structural changes, contributed significantly to economic growth in the Eastern region. As a result, the divergence of the economic growth rate and income inequality between the two regions have further widened in the past 18 years.

The increased inter-regional economic disparity reveals that the economic boom of the Eastern region has not diffused effectively to the Western region. The diffusion of growth (the 'trickle down effect') from the Eastern region to the Western region has not happened or has not been empirically evidenced. The major reason for the deficiency of the diffusion of growth is the lack of effective inter-regional industrial linkages and the lack of an integrated and well-functioning domestic market.

Under the open-door policy, the Eastern region has been encouraged to be more involved in the international markets for both the export of products and the import of inputs. This development strategy was formally confirmed at the 13th Congress of the Chinese Communist Party in 1987. As a result, the economy of the Eastern region become more foreign market-oriented. This is particularly the case in Guangdong and Fujian provinces. Their economic integration with Hong Kong and Taiwan has developed rapidly in the past ten years and is an important characteristic and major cause of economic development of the Southeast Region of China. However, as the coastal region has increasingly shifted emphasis to overseas markets, the economic linkages between the coastal region and inland regions including the Western region have weakened.

In addition, the less-developed domestic market has worked against developing effective regional economic linkages. Since the early 1980s, the Chinese economy has been undergoing a transformation from the traditional centrally-planned system to a market economy. The market-oriented reforms are more progressive and far-reaching in the Eastern region than in the Western region. The regional imbalance in progress towards economic reforms and opening has impeded the formation and development of the integrated domestic market and restrained the economic linkages and cooperation between the two regions. Even at the current stage, regional market segments still prevail in China. As a result, economic growth in the Eastern region can not effectively diffuse to the Western region, as the 'dualism' theory predicts. Consequently, the inter-regional economic disparity and income inequality has widened.

7.7 Conclusions and Policy Implications

In the reform era since 1979, the Chinese economy has experienced rapid growth and a widening disparity in economic indicators between regions. Many factors have contributed to the regional divergence in economic growth and increased income inequality. Industrial structure and resource conditions, coast-oriented foreign investment and economic openness, regional development policy with emphasis on the Eastern region, the unbalanced growth of exports and direct foreign investment, rural industrial development, and domestic capital flows from the Western region to the

Eastern region have all contributed to inter-regional differences in economic growth and income per capita.

This chapter has investigated the factors behind the divergent growth of the two regions, especially structural factors, direct foreign investment and economic openness. Using production function analysis, this study has found that the higher rate of domestic investment, enormous inflow of direct foreign investment, and greater openness are the major reasons for the higher economic growth rate of the Eastern region compared to the Western region. Without a large amount of DFI inflow and greater openness, the economic growth rate of the Eastern region would not be significantly higher than that of the Western region. This suggests that foreign investment, exports and greater involvement in international market have promoted capital formation and productivity, which are the major determinants of a higher rate of economic growth in the Eastern region. As a result of divergent growth rates, the economic gap and income inequality between the Eastern and Western regions have been further widened in the reform era.

The disparity which has arisen is also closely related to the regional policy of the Chinese government. In essence, the regional policy throughout the reform era was based on regional comparative advantages. Under this policy, the coastal region, due to its superior geographic location and factor endowment, has been given a pivotal role as 'growth pole' or 'engine of growth'. To promote the growth of this region, the central government implemented special economic policies and committed a large amount of capital to improve infrastructure. As a result, the investment environment in the coastal region was more attractive than in the inland. Domestic and foreign capital therefore flowed into the coastal region, accelerating its regional economic growth. However, due to the lack of effective inter-regional industrial linkages and a well-functioning domestic market, the economic boom in the coastal region has not noticeably spread to the inland, further resulting in a widening inter-regional economic disparity and income inequality.

Some implications can be drawn from the Chinese experience. First, inter-regional disparity is unavoidable in the process of economic development, especially for a developing country. Since the conditions for development in each region are different, the regional policy based on comparative advantages tends to reinforce the existing regional difference. Thus, regional 'equality' and 'efficiency' of development are two primary issues faced by the government of a developing country. In the initial stage of development, a government may place priority on a coastal (well established) region which can lead to rapid growth of the national economy as a whole.

Second, a government should pay particular attention to the growing inter-regional disparity. In order to facilitate the diffusion of growth from the booming region to a relatively stagnant region, inter-regional economic linkages should be promoted through a series of policy tools and encouragement of economic cooperation across different regions. The artificial barriers to inter-regional economic linkage and integration should be removed, and in the Chinese context, some regulations which isolate the SEZs or other coastal areas from the inland regions should be eliminated. Furthermore, bilateral trade, investments and other linkages between the coastal and the inland regions should be promoted, rather than the coastal region being encouraged to rely solely on foreign markets.

Another important point for a less-developed market economy or for one undergoing the transformation from a central planning system to a market economy, is the enhancement of the market mechanism. As a well-developed market can allocate resources efficiently, the overall efficiency of an economy can be significantly improved. However, since market forces may fail to facilitate a balanced growth and may even result in regional inequality in the first stage of development, a suitable regional policy is critical. The government should foster market efficiency and domestic market integration and promote economic linkages and cooperation between regions. Therefore, the role of the government in achieving a balanced growth is important.

Third, as the coast-oriented open policy has tended to widen the economic gap between the Western and the Eastern regions, it is necessary for the Chinese government to standardize the open policy in all regions and to eliminate the existing regional discrimination in foreign trade autonomy. The open policy should be extended as much as possible to the vast inland regions where a huge potential for development exists. This would enable the Western region to improve economic conditions and explore its great development potential.

Finally, improvement in transportation and communication facilities is essential for enhancing inter-regional economic linkages and cooperation. The central government should allocate funds to finance some important transportation projects, and should also encourage local capital to invest in infrastructure so as to improve the local investment environment. To this end, some special programs stimulating investment in the Western region are necessary.

Notes

1　China consists of three macro-regions. The Eastern (coastal) region includes eleven provinces and cities: Guangdong, Fujian, Jiangsu, Zhejiang, Shanghai, Shandong, Hebei, Beijing, Tianjin, Liaoning and Hainan. The Central region contains ten provinces: Heilongjiang, Jilin, Inner Mongolia, Shanxi, Henan, Hubei, Hunan, Jiangxi, Anhui and Guangxi. The Western (far inland) region includes nine provinces: Shaanxi, Sichuan, Gansu, Qinghai, Xinjiang, Ningxia, Guizhou, Yunnan, and Tibet. Although this study includes brief references to the Central region, it focuses on the Eastern and Western regions

2　The recent publication of *The East Asian Miracle* (1993) by the World Bank is a testimony of this view. In this publication, East Asia comprises all the low- and middle-income economies of East and Southeast Asia.

3　It is a common practice to add a time trend 'T' in the production function to measure the rate of technological progress over time. However, in the current case, it has been found that the addition of 'T' results in a biased coefficient for DFI because 'T' and DFI are highly correlated with each other (0.81 for the Eastern region and 0.55 for the Western region). Therefore, time trend is removed from the production function in this study, and thus a constant technological change is assumed. The effects of factors other than direct inputs (capital and labor) on output, including technology, management and economic structure, can be reflected in the value of the constant term 'A' in the production function. This is because 'A' fixes the position of the isoquant and therefore reflects the influence of technical progress on the movement of the isoquant (see Heathfield, 1971, for a detailed discussion).

4　The Cobb-Douglas production function assumes a constant elasticity of substitution of unity, which means that a 1 percent change in the factor's relative prices will bring about a 1 percent change in factor proportions. The different growth rate of factors, as listed here, imply that the relative prices of factors also change at the same rate. In a distorted market like China, relative prices of factors (capital and labor) are difficult to estimate.

8 DFI, Foreign Trade and Transfer Pricing

8.1 Introduction

DFI contributes to capital formation, productivity growth and economic development in the host country through transferring a package of capital, technology and management skills from investing countries to the host country. However, the positive impact of DFI on foreign trade remains unclear. The key issue is whether DFI creates or replaces foreign trade? At the macro level, Kojima (1973, 1975, 1982) classifies DFI into two categories: trade-creating and trade-substituting. The first type of DFI creates more trade opportunities between investing and host countries, and therefore, complements foreign trade, whereas the second type of DFI replaces foreign trade or destroys trade conditions. At the microeconomic level, the role of multinational corporations (MNCs) in international trade, especially the effects of MNCs' intra-firm trade on the trade balance of the host country, are open to debate among academics and policy-makers.

This chapter investigates the impact of DFI on foreign trade in the case of China, using both a macro and micro approach. It briefly discusses the theories of the impact of direct foreign investment on foreign trade, and then presents an empirical investigation of the influence of DFI on China's exports, imports and trade balance, providing an empirical test for Kojima's hypothesis about DFI 'trade-creation' and 'trade-replacement'. Finally, this chapter investigates transfer pricing by MNCs in the case of China, probing their motivations and latitude to practice transfer pricing in Chinese circumstances and examining empirical evidence.

8.2 Theoretical Analysis of the Impact of DFI on International Trade

How international investment affects trade, and what role multinational corporations may play in international trade are important issues which can be studied at both the macro and micro levels. At the macroeconomic level, the activities of multinational corporations change factor endowments which underlie international trade. MNCs have increased international capital mobility through direct investment in different countries which can be expected to alter the factor proportion (capital-labor ratio) in the home and host countries. Although it may be true that sometimes a company may establish a subsidiary in a particular country using capital drawn from the host country's domestic market, the basic direction of capital flows from capital abundant country to capital scarce country remain virtually the same.

DFI also involves the transfer of technology, management know-how, entrepreneurial skills, and labor training. In essence, this is the transfer of particular form of capital - human capital. This will affect both the quantity and quality of factors and their combinations in relevant countries, resulting in changes in factor productivity and comparative cost advantage between products. Such a *dynamic change in comparative advantage*, as labeled by Kojima (1973, 1975), will inevitably affect international trade both in structure and in direction.

At the microeconomic level, a significant part of international trade is carried by and within multinational corporations. Intra-firm trade and transfer pricing by MNCs also change the pattern and structure of international trade and effectively influence the distribution of income and trade gains in relevant countries.

To date there is still no a broadly accepted composite theory in which the international trade theories have been integrated with international investment theories. Part of the difficulty lies in relating international direct investment with international trade outcomes.

8.2.1 Foreign Investment Substitutes for or Complements to Trade: A Macro Approach

The traditional theory of international trade, as represented by the Heckscher-Ohlin-Samuelson (H-O-S) model, emphasizes the differences in factor endowments between different countries. According to the theory, differences in resource endowments and factor proportions are the determinants of international trade. A country should specialize in the production of the good which it possess comparative cost advantage. By producing and exporting its comparatively advantaged product, the country will benefit from international trade. All trade partner countries will be better off from a net trade gain.

It is a central proposition of the H-O-S model that factor movements are, at least to some extent, a substitute for trade, and vice versa. Ohlin (1933) stressed that trade on its own would *tend* to equalize factor prices; and factor movements would tend to bring closer factor prices in the two countries. However, due to the existence of transport costs, trade tariffs and differences in production functions, trade and factor movements are not perfect substitutes.

Based on Ohlin's work, Mundell (1957) has explored the special case where trade and factor movements are perfect substitutes. Under the restrictive assumptions of zero transport costs, identical production functions for each good in the two countries, and different factor intensities being required by different products, he argues that free trade tends to result in commodity price equalization, in a tendency toward *factor-price equalization* in different countries. It is equally true that perfect mobility of factors results in factor-price equalization and, even when commodity movements cannot take place, in a tendency toward *commodity-price equalization*. He argues that under the condition of perfect capital mobility, all trade will cease since the basis of trade - differences in factor proportions between countries will no longer exit. Thus, trade and factor movements are substitutes and replacing for each other. Alternatively, the capital abundant country may export its capital instead, to the point where it ceases to be relatively well-endowed with capital (Corden, 1974). Therefore, international capital movements would eliminate commodity movements. Clearly, Mundell's conclusion is just the logical result of two extreme cases, as soon as the restrictive assumptions are relaxed, the conclusion will disappear.

By contrast, Purvis (1972) proposed a model where international capital flows are complements to commodity trade. He defines complements as follows: 'A sufficient condition for complementary is that the initial capital outflow generates an excess demand for imports and an excess supply of exportables at constant term of trade.' If foreign investment creates and /or expands the opportunity to import one product and to export the other product, it is complementary to product trade. On the other hand, if initial capital outflow reduces or eliminates the opportunity to import one product and to export the other product, it will substitute for product trade and is thus trade-destroying. Based on his analysis of the effects of different production functions for two goods between two countries and their comparative productivity, he concludes that the condition for capital movements to be complements to product trade is that the overall output value of the two goods in the two countries taken together must increase, otherwise trade would be partly eliminated and welfare for the two countries would not rise. However, Purvis's argument does not clarify in what conditions, foreign investment complements and promote product trade.

Kojima (1973, 1975 and 1982) further developed the Mundell and Purvis models, and specified the conditions for foreign investment to be complementary to or substituting for commodity trade. He has played a pioneering role in developing a systematic macroeconomic approach to direct foreign investment, and in incorporating DFI with international trade theory. As a starting point, he distinguishes direct foreign investment from international money capital movements. Not only transfer capital, but also DFI transplant superior production technology through training of labor, transfer of management and marketing know-how, from advanced industrial countries to less developed countries. As a result of technology transfer, superior production functions can be transferred from investing countries to the host country. In addition, through the activities of subsidiaries and joint ventures in particular industries, foreign investing companies can spread over the industries in the host country and play a role in training local laborers, engineers and managers, ultimately improves the productivity of local firms.

In what type of industries can DFI easily transfer technology and improve production functions in the host country to eventually create more trade opportunities? Kojima argues that if DFI flows into industries in which the host country has comparative advantage, it tends to improve productivity of the host economy and therefore stimulates more exports. This is because the smaller the technological gap between the investing and host countries, the easier it becomes to transfer technology and improve productivity in the latter. In terms of developing host countries, DFI flowed into labor-intensive industries are largely trade-creating. DFI flowed into capital-intensive industries where the host country is comparatively-disadvantaged, is trade-replacing, or trade-destroying as products produced by such investments are largely import-substituted, resulting in a decrease in trade between the investing and host countries.

On the part of the investing country, if outward investments come from its comparatively disadvantaged industries, e.g. labor-intensive or resources-based investments, they tend to be trade-creating; if outward investments come from its comparatively advanced industries, they tend to be trade-replacing since such investments do not fit with the host country's comparative advantage, and eventually reduce the total output of the two countries and their trade volume.

Some authors (e.g. Arndt, 1974; Geroski, 1979; Mason, 1980; Buckley 1983, Lee, 1984 1985; and Dunning, 1985) cast doubt on the validity of Kojima's hypothesis. They argue that international investments made by multinational corporations may be diversified in various industries including capital/technology intensive and labor-intensive industries, depending on firms' competitive advantage in the host country's market. The resultant net impact on foreign trade is uncertain. In addition, Kojima's theory may fail to explain the 'double-way' international investment flows between industrialized countries (e.g. between the U.S. and Europe), although it may be applied to international investments flowing from industrialized countries to developing countries.

8.2.2   Intra-firm Trade of MNCs: A Micro Approach

The activities of MNCs have an important impact on international trade. International direct investments made by MNCs reallocates globally economic resources and the productive capacities according to relative costs of production in different countries. This is expected to bring about a dynamic change in comparative advantages in relevant countries, leading to changes in the structure and pattern of international trade. In addition, global activities of production, marketing and research and development (R & D) by MNCs build world-wide linkages, which in turn facilitate transfer of technologies and quicker flows of information. This tends to stimulate the flow of commodities.

Furthermore, horizontal and vertical integration developed by MNCs through their international production activities require, and also promote, international circulation of components, parts and other intermediate products between different countries. This would certainly increase the volume of international trade. In terms of the format of DFI, joint ventures with local firms can significantly improve the host country's access to international markets.

DFI can also affect the host country's economy through industrial linkages especially backward linkages effects. Foreign subsidiaries in the host country buy locally made intermediate inputs such as parts and components, for production of exportables. This will stimulate exports from domestic firms. The exports induced by DFI can constitute a considerable part of total exports from the host country. Finally, intra-firm trade within a MNC's operating network becomes increasingly important in international trade. It is estimated that at least 25 percent of international trade is carried by MNCs through intra-firm trade (United Nations, 1992). In addition to direct intra-firm trade within MNCs, a substantial volume of trade is likely to be conducted by MNCs (Gray, 1993).

Therefore, knowledge of intra-firm trade taking place within a MNC's network is crucial to a full understanding of the effects of MNCs on international trade. The growth of intra-firm trade not only increases the volume of international trade, but also changes the pattern of international trade and the distribution of income and gains from trade. Unlike trade between independent parties - so-called 'arm's length trade', intra-firm trade is controlled by the same company. To avoid taxation, tariff and business risks, MNCs may manipulate transfer prices for such intra-firm trade. Through overinvoicing or underinvoicing exports or imports, MNCs shift profits out of the host country and minimize business risks.

As a result, host countries, especially developing host countries, may suffer from income loss. Furthermore, manipulative transfer pricing is likely to result in a worsening balance of trade for a developing host country as exports are underpriced and imports are overpriced by MNCs. Since domestic firms in developing countries lack expertise and an appropriate marketing network, MNCs not only directly control intra-firm trade between their units in different countries, but largely control the foreign trade of local firms. The coverage of transfer prices, therefore, not only include the direct exports and imports of their subsidiaries, but also to some extent the trade of domestic firms. The propensity and latitude for a MNC to pursue transfer pricing is dependent on taxation, tariff, economic policies and business risks in relevant countries.

8.3   DFI and Foreign Trade of China: An Empirical Investigation

The theoretical discussion above indicates that DFI may play an important role in the foreign trade of the host country by creating more trade opportunities. It is also possible that DFI may partly replace international trade, depending on how DFI changes the existing comparative advantages in the home and host countries. If DFI fit with and strengthens the host country's comparative advantage (comparative disadvantage for the home country), it would increase foreign trade. A number of empirical studies (e.g. Hone, 1974; Helleiner, 1973; Rana, 1985) on the impact of DFI on the host country's foreign trade, have concluded that DFI contributes to the export growth of the host country. In this section, an empirical investigation of this issue will be developed in the case of China.

Since China began opening to the outside world in 1979, exports have grown rapidly. During the period from 1979 to 1996, Chinese exports increased by 16.4 percent annually. As a result, the degree of the openness of the Chinese economy, which can be measured by the sum of exports (X) and imports (M) as a share of GDP, i.e. (X+M)/GDP, rose from 10.3 percent in 1979 to 22.1 percent in 1997[1] (SSB, SYOC 1997).

Among the factors promoting export growth, DFI has played an important role, with foreign-invested enterprises (FIEs) being the most active players in the export expansion drive. During the period 1981 to 1995, exports by FIEs grew at an annual rate of 63.3 percent, with the value of exports increasing from US$32 million to US$468,80 million. As a result, the share of FIEs in the total exports of China increased from 0.1 percent in 1981 to 31.5 percent in 1995 (see Table 8.1). The rapid expansion of FIEs' exports led to the phenomenal growth of China's exports.

In the coastal region, the contribution of FIEs to the exports is even more significant. As shown in Table 8.1, FIEs provided nearly half of the total exports of Guangdong Province. In 1995, the share of FIEs in total provincial exports reached 43.6 percent. For the period 1986 to 1995, the exports of FIEs in Guangdong increased by US$25,370 million, representing 44.3 percent of the total increase in the provincial exports.

Due to the FIEs' driving force, Guangdong's exports grew at an annual average rate of 30.1 percent during the period 1986 to 1995. This is a significantly higher rate than the national average growth rate of 17 percent. Similarly, in other coastal provinces FIEs have also contributed considerably to exports. For instance, in Fujian and Tianjin, FIEs provided 43.7 percent and 44.8 percent respectively of the total exports in 1995.[2] In Shanghai, FIEs provided 30 percent of the total exports in the same year. Therefore, the impact of DFI on the exports of China (especially in the coastal region) has become of major importance.

In terms of category of commodities, FIEs' exports are concentrated in labor-intensive products including textile, garments, shoes and other fabric products, food, beverage & wine, handicraft articles, toys, TV receivers & sound equipment, clocks and watches, bicycles and parts, plastic articles, household electrical products, sports goods, general metalware, office supplies and other light manufactured goods. In 1993, these commodities represented 81.6 percent of the total exports of FIEs (calculated from Table 9.2).

As outlined in Table 8.2, FIEs have become the dominant exporter for a number of labor-intensive products, which comprise primary commodities exported from China. For example, in 1993 FIEs provided over half of the total value of some of the major Chinese exports such as TV and sound equipment (53.5 percent), clocks and watches (53.6 percent), plastic products (56.9 percent), office supplies (55.1 percent), sports goods (60.8 percent) and industrial tools and instruments (53 percent). In some capital /technology intensive products, FIEs have provided an increasing volume of the exports. These products include machinery and equipment, transport equipment, and chemical products. In 1993, FIEs accounted for 25 percent, 30 percent and 15 percent respectively of these three categories of exports. However, these capital/technology intensive exports accounted for a small share of total exports of FIEs, being less than 20 percent.

The structure of exports by foreign-invested enterprises, as discussed above, provides some empirical support for Kojima's hypothesis that if foreign investments flow into industries in which the host country has comparative advantage, they tend to be trade-creating, and thus strengthen exports from the host country. By investing in labor-intensive industries, foreign companies can combine their advantages in capital, technologies, and management and marketing skills with Chinese labor resource, creating new competitive advantage in international markets. In addition, foreign investments in the comparatively advantaged industries of China improve the efficiency of resource allocation at the macro level as a large labor force would otherwise be unemployed or underemployed if foreign investment rather flowed into capital/technology intensive industries.

At the micro level, foreign investment in China's comparatively advantaged industries is likely to increase factor productivity since factor proportions have been changed (capital/labor ratio rises) by foreign investment in China. This also creates a net gain for investing countries given that labor-intensive industries are their comparatively disadvantaged industries. Capital outflows from these industries tend to increase return to capital for existing firms in these industries. Decrease in supply of these products in the investing country's domestic market tend to stimulate imports from the developing host country. Therefore, further trade opportunities tend to be created by DFI of this type.

Another type of DFI in China is domestic-market oriented. China has a large domestic market with 1.2 billion consumers. With strong economic growth, the domestic markets have grown rapidly attracting many foreign companies to invest in China. In large part, they have targeted the domestic markets both for consumer goods and producer goods with a wide variety of capital/technology contents. These type of investments are in essence trade-substituting since the products and services produced by foreign-invested enterprises are sold in the domestic markets as substitutes for imports. Therefore, this type of DFI tends to worsen the trade balance of China because intermediate inputs used in FIEs are largely imported from overseas especially from MNCs' home countries, and their products are sold in the Chinese domestic markets.

As shown in Table 8.1, over the period 1981 to 1995, the foreign trade of FIEs was not balanced, with the imports of FIEs constantly exceeding their exports. Only once (in 1983), were FIEs' exports slightly larger than their imports. As a result, the trade deficit of FIEs has experienced a strong growth, from US\$26 million in 1980 to US\$16,060 million in 1995, affecting China's trade balance negatively. Prior to 1989 FIEs' trade deficit exacerbated the existing trade deficit of China. It was an important factor behind the large trade deficits of China for that period. However, in the most recent six years (1990-95), the increased trade deficits of FIEs have largely offset the trade surplus created by the Chinese domestic sectors.

A proper assessment of the net impact of FIEs' trade deficit on the Chinese economy requires an analysis of the reasons for the trade deficits. By definition, FIEs' trade deficits result from the value of their imports exceeding the value of their exports. There are three factors contributing to FIEs' trade deficits. First, FIEs import machinery and equipment as investment goods. These capital goods have accounted for a sizable proportion of FIEs' total imports in the past few years. For instance, in 1993, 39.5 percent of FIEs' imports were machinery and equipment as investment goods in FIEs.[3] If this portion of imports are excluded from FIEs' total imports, their foreign trade has been balanced or has had a slight surplus since 1990. The importation of investment goods, especially advanced machinery and equipment, assists fixed capital investment in China, and promotes economic growth in the long run.

Second, FIEs import the inputs of production including raw materials, and other intermediate products. This is the major cause of trade deficits for FIEs who rely on inputs imported from overseas and sell their products in the domestic market. Such trade deficits may effectively affect the host economy. They represent a net leakage of aggregate demand from the importing country, thereby negatively affecting the economy. On the other hand, if FIEs' products, especially capital goods, are used as substitutes for imports, China would save foreign exchange.

Table 8.1 FIE's Effect on the Foreign Trade of China and Guangdong Province[1] **(100 million US$)**

	China								Guangdong							
	Exports			Imports			Trade Balance		Exports			Imports			Trade Balance	
	Total	FIEs	%	Total	FIEs	%	Total	FIEs	Total	FIEs	%	Total	FIEs	%	Total	FIEs
1980	181.2	0.1		200.2	0.3	0.2	-19.0	-0.3	22.1			3.6			18.5	
1981	220.1	0.3	0.1	220.2	1.1	0.5	-0.1	-0.8	24.2			6.7			17.5	
1982	223.2	0.5	0.2	192.9	2.8	1.4	30.3	-2.3	22.7			8.0			14.7	
1983	222.3	3.3	1.5	213.9	2.9	1.3	8.4	0.4	24.0			9.4			14.6	
1984	261.4	0.7	0.3	274.1	4.0	1.5	-12.7	-3.3	25.2	0.7	2.9	12.1			13.1	
1985	273.5	3.0	1.1	422.5	20.6	4.9	-149.0	-17.6	29.5	2.2	7.3	24.3	3.4	11.9	5.2	-1.2
1986	309.4	5.8	1.9	429.0	24.3	5.7	-119.6	-18.5	42.5	3.9	9.2	25.6	8.5	32.3	16.9	-4.6
1987	394.4	12.1	3.1	432.2	31.2	7.2	-37.8	-19.1	54.4	6.2	11.4	36.2	8.0	22.0	18.2	-1.8
1988	475.2	24.6	5.2	522.8	57.5	11.0	-47.6	-32.9	74.8	12.0	16.1	51.1	11.3	22.1	23.7	0.7
1989	525.4	49.1	9.4	591.4	88.0	14.9	-66.0	-38.9	81.7	22.8	30.2	48.3	19.5	40.4	33.4	3.3
1990	620.9	78.1	12.6	533.5	123.1	23.1	87.4	-45.0	105.6	37.2	37.6	57.5	33.0	57.4	48.1	4.2
1991	719.1	120.5	16.8	637.9	169.1	26.5	81.2	-48.6	136.9	53.3	41.6	85.1	45.1	53.0	51.8	8.2
1992	850.0	173.6	20.4	806.1	263.9	32.8	43.9	-90.3	184.4	81.6	44.3	111.8	60.2	53.8	72.6	21.4
1993	917.6	252.4	27.5	1039.5	418.3	40.2	-121.9	-165.9	376.0	143.7	38.2	431.2	198.0	45.9	-55.2	-54.3
1994	1210.1	347.1	28.7	1156.1	529.3	45.8	54.0	-182.2	532.7	198.4	37.2	477.9	253.6	53.1	54.8	-55.2
1995	1487.7	468.8	31.5	1320.8	629.4	47.7	166.9	-160.6	590.5	257.6	43.6	494.8	274.4	55.5	95.7	-16.8

Note: 1. The figures in this table are based on the China's Customs Statistics.

Source: SSB, *China's Foreign Economic Statistics 1979-1991, 1994 and 1996*; and *Statistical Yearbook of China 1996*. The Statistical Bureau of Guangdong Province, *Guangdong Sheng Duiwai Jingji Guanxi Tongji Zilao (Statistical Materials of Foreign Economic Relations of Guangdong Province), 1987-1992,* propelled by.

Table 8.2 FIE's Contribution to the Exports of China by Commodity   (Unit: million US$)

Commodities	1988			1990			1992			1993		
	FIEs[1]	Total	%	FIEs	Total	%	FIEs	Total	%	FIEs	Total	%
Total	1746	40640	4.3	6021	62091	9.7	15591	84940	18.4	23390	91763	25.5
Food, beverage & wine	151	3602	4.2	304	4327	7.0	780	5654	13.8	1344	6152	21.9
Textile fibbers & fabric	350	7431	4.7	825	6235	13.2	2031	10465	19.4	2979	12242	24.0
Garments, shoes & other fabric	240	4270	5.6	1359	6498	20.9	3583	12385	28.9	5696	17656	32.0
Household electrical appliances	57	509	11.1	247	873	28.2	563	1728	32.6	803	3179	25.3
TV receivers, sound equipment	270	744	36.3	712	1550	45.9	1018	1887	53.9	982	1836	53.5
Machinery and equipment	23	997	2.3	134	1877	7.1	349	2592	13.5	702	2839	25.0
Bicycles and spare parts	4	148	2.4	116	308	37.6	184	554	33.3	229	537	42.7
Plastic articles	59	280	20.9	213	447	47.6	526	823	63.9	847	1490	56.9
Handicraft article	69	736	9.3	224	899	24.9	708	1689	41.9	1021	2084	49.1
Toys	55	268	20.7	134	314	42.8	351	660	53.1	535	1512	35.4
Chemicals	55	1795	3.1	138	2627	5.3	351	3746	9.4	590	3854	15.0
Transport equipment	4	475	0.8	50	857	5.9	349	1545	22.6	551	1843	30.0
Clocks and watches	37	131	28.3	122	299	40.9	162	295	55.0	226	421	53.6
Metallic products	62	2201	2.8	172	2421	7.1	395	3685	10.7	639	3998	16.0
Tools and Instruments	81	553	14.7	383	1422	26.9	1394	2603	53.5	1818	3440	53.0
Sports goods	2	171	1.2	124	345	36.1	464	838	55.3	750	1234	60.8
Office supplies	16	38	43.3	35	67	52.2	57	103	55.1	74	135	55.1
General metalware	12	161	7.2	45	245	18.3	141	553	25.4	239	742	32.2
Other	200	16130	1.24	684	30481	2.3	2187	33134	6.6	3359	26570	13.0

Note: 1. The figures of FIE's exports and China's total exports in this table are based on the MOFTEC's statistics, which are some
times not identical to the statistics published by the Customs.

Source: SSB, *Foreign Economic Relations Statistics*, 1979-91, 1994 and 1996.

Third, transfer pricing by foreign investors (MNCs) for FIEs' exports and imports is another reason for the persistent trade deficits of FIEs. As FIEs' foreign trade is largely controlled by MNCs, their exports tend to be undervalued, with imports being overvalued. This results in trade deficits for there enterprises and in income loss for China. The above factors simultaneously contribute to the trade deficits of FIEs, with the net impact on the Chinese economy depending on the relative strength of each factor.

From a perspective of development, the impact of DFI on the host country's net exports can be considered into two stages. In the first stage (e.g. the first 15 years) it could be negative, but in the second stage it turns positive. This is because in the initial stage, FIEs highly rely on imported inputs including equipment, intermediate products and raw materials. The low quality and unstable supply of these inputs by local firms are the major reasons for a lower local content in total inputs used in FIEs. However, as the quality and supply system of locally made inputs improves, FIEs will use more locally made inputs. Accordingly, the ratio of imported inputs to total inputs will decline. The input localization policy pursued by the Chinese government also promotes this trend. Therefore, the effect of DFI on the net exports of China is expected to become positive in the near future.

8.4 Transfer Pricing and FIEs' Trade

Transfer prices are the prices that MNCs set for intra-firm exports and imports across national boundaries. As these prices are set by MNCs for transactions between their related units, they are different from the arm's-length market prices used for transactions with unrelated units. The motivation of MNCs to manipulate transfer pricing is to maximize their global profit and minimize the total costs by avoiding or reducing taxes and tariffs in their home countries and in host countries where their subsidiaries operate. The principal tools utilized by MNCs to manipulate transfer pricing are underinvoicing (i.e. underpricing) the exports and overinvoicing (i.e. overpricing) the imports of the host country.

Many factors influence MNC's transfer pricing manipulation. These include tax rates and tariffs at home and in host countries, import or export restrictions, foreign exchange control, restrictions on the repatriation of profits, joint venture partners' capability to manage and influence the pricing policy and practice and taxation (or Customs) authority's expertise and ability to detect and prevent transfer pricing. In general, a large international

differential in taxes and tariffs would give rise to a high propensity for
MNCs to use transfer prices to reduce tax payments and tariff duties in
relevant countries. In addition, business environments such as restrictions
and regulations on imports, exports and foreign exchange control, can also
stimulate MNCs' incentives to manipulate transfer pricing to avoid business
risks and transaction costs.

For a particular host country, a set of factors may simultaneously
influence MNCs' transfer pricing behavior. Some factors may strengthen
MNCs' motivation to employ transfer prices while others could limit their
capacity to practice transfer pricing. Therefore, it is necessary to analyze the
factors stimulating MNCs' motivation to use transfer prices and the factors
limiting the scope and latitude of MNCs' transfer pricing.

8.4.1 MNCs' Motivation to Use Transfer Prices

Taxation, tariffs, business risks and transaction costs are the major
determinants of MNCs' motivation to manipulate transfer pricing. In terms
of taxes and tariffs affecting DFI, China is favorably compared to many
other developing countries. Under China's FIE income tax law, all FIEs
including joint ventures and foreign wholly-owned enterprises are granted a
package of special preferential tax treatments. These treatments include a
two-year tax holiday starting at the first profit-making year, and a 50 percent
tax reduction for the following three years.

For FIEs investing in particular industries and locations, the tax
treatments are even more favorable. For example, foreign investments in
'low-profit industries' (like agriculture and forestry) and high technology
industries are eligible for a two-year extension of the tax holiday. DFI in the
Special Economic Zones (SEZs) and Economic and Technology
Development Areas (ETDAs) in the open coastal cities are eligible for a
lower income tax rate after the tax holiday and a longer tax concession
period. The income of FIEs in the SEZs and ETDAs is subject to a tax rate
of 15 percent compared to 33 percent in other regions.

In addition, provinces and cities in various regions provide DFI with
more generous tax treatments (e.g. a longer tax holiday and tax reduction
period) and other forms of preferential treatments such as lower land using
fees. Therefore, within the tax holiday and tax reduction period, MNCs do
not have particular incentive to avoid tax by using transfer pricing. The
MNCs' motivation for *tax-induced transfer pricing* in China, especially
during the periods of tax holiday and tax reduction, should be low. After the

period of the tax holiday and tax reduction, FIEs are subject to the standard corporate income tax rate of 33 percent. This is close to the corporate tax rates in many other countries, but is significantly higher than that in Hong Kong (only 15 percent). Thus, the tax-induced transfer pricing motivation would become quite high.

The current tariffs for FIEs' exports and imports are quite preferential, with FIEs' exports effectively being tariff free. Imports by FIEs which are used as investment goods and for producing exports are also tariff free.[4] Only imports (by FIEs) used to produce goods to be sold in the domestic market are subject to some import tariffs. Therefore, tariffs applied to export-oriented FIEs in China are generous, and generate no particular incentive for MNCs to manipulate transfer pricing.

Nevertheless, the business environment in China may strengthen MNCs' motivation to use transfer prices to minimize business risks, transaction costs and market unpredictability. It is a tendency in developing countries to impose more constraints on the operations of MNCs than is the case in developed countries. In the case of China, although the government has made great efforts to establish a legal framework for DFI and to liberalize the economy, there are still some legal and institutional factors negatively affecting the operation of FIEs.

A major concern of foreign firms investing in China is the lack of well-defined laws on property rights and foreign investment protection. In many cases, the property rights in enterprises are not legally clear. Although a number of laws (like the corporation law, the bankruptcy law and the trade mark law) have been passed since the early 1980s, these laws are far from being fully implemented. As a result of the poor legal environment, foreign investors tend to transfer their income to low-risk countries by using transfer pricing.

In addition, a number of restrictive regulations such as foreign exchange control, withholding tax on profit repatriation, and restraints on FIEs' access to domestic markets, increase business risks and relevant transaction costs. These restrictions motivate MNCs to use transfer prices in order to shift their income to countries with minimal restrictions. Moreover, the inconvertibility of the Chinese currency is another factor inciting MNCs to keep their export income out of China. Therefore, the *risk-avoiding* type of transfer pricing tends to dominate the motivation of MNC to practice transfer prices in the case of China.

Another important factor that deserves special attention is the ownership structure of FIEs in China. Joint ventures including equity joint ventures and cooperative joint ventures are the principal forms of FIEs. By law, both sides in Sino-foreign joint ventures share profits, costs and management. In practice, however, as the interests and management goals differ between the two sides, each side is motivated to maximize its own interest and benefit. To reduce the profits accruing to local Chinese partners and to increase their own real profit share, MNCs have a high propensity to manipulate transfer pricing to shift profits to their parent companies or other subsidiaries located in other countries with low-taxation and risks.

8.4.2 *The Scope and Latitude of MNCs' Transfer Pricing*

The propensity and motivation of MNCs to use transfer pricing is quite high in China. The extent to which transfer pricing is practiced is determined by many factors including the anti-transfer pricing regulation of the host country government, the taxation and customs authorities' expertise on arm's-length market prices and their capability to detect and prevent MNCs' manipulation of transfer pricing, and the local joint venture partners' ability to check and prohibit MNCs' transfer pricing practice. In general, the governments of developing countries are poorly equipped in comparison with developed countries to thwart the attempts of MNCs to manipulate transfer pricing (Plasschaert, 1985). In the case of China, there is lack of necessary regulations against transfer pricing manipulated by MNCs. The government departments, especially the taxation department and the customs authority, have recognized this problem (transfer pricing) prevailing in FIEs, but to date they do not have a special regulation against transfer pricing. This could be attributed to difficulties in investigating MNCs' transfer pricing practice and setting the arm's-length market prices for a large number of commodities traded by FIEs internationally.

At the firm's level, the export and import business in many joint ventures are controlled by foreign investors (i.e. MNCs). The Chinese partners in general lack professional knowledge of international trade. This excludes them from FIEs' pricing decisions for exports and imports, thereby providing a ideal environment for MNCs to perform transfer pricing at the cost of Chinese partners' profits. In some cases, Chinese partners may participate in the decision-making process for imports and exports, but the managing power of FIEs is generally controlled by foreign partners arising from the particular management structure of joint ventures. As a general rule, in joint ventures foreign investors or their representatives hold the position of general manager (or chief executive) and run the enterprises at their discretion. The Chinese side takes the position of chairman (or president) of the board of directors, who are generally government officials without professional qualifications and appropriate management skills. Therefore, the existing management structure of joint ventures leaves a large latitude for MNCs to manipulate transfer pricing, although it can be said that the existence of Chinese partners tends to impede MNCs transfer pricing practice and curb their ability to underinvoicing exports and overinvoicing imports.

Finally, different types of DFI may vary in their scope to use transfer prices, depending on the market orientations of their investment. In terms of market orientation, DFI in China can be classified into two categories: export-oriented and domestic market-oriented. For the first category of DFI, products produced by FIEs are primarily exported overseas, and inputs can be obtained from either domestic suppliers or importing. The current free-tariff regime for FIEs' exports and imported inputs used for production of exports, facilitates FIEs in exporting and importing. Thus, this tariff regime is likely to be conducive to overinvoicing of imports and underinvoicing of exports. With the free-tariff regime, MNCs are more able to manipulate transfer pricing and consequently, a larger share of FIEs' income may be transferred to MNCs' system.

In contrast, for DFI oriented toward Chinese domestic markets, MNCs have a lower propensity and capability to use transfer prices. Essentially trade between FIEs and non-affiliated enterprises in Chinese domestic market use market prices. However, if FIEs use imported inputs, MNCs still have the incentive to overinvoice imported materials from their own overseas subsidiaries. This results in a shift of some profit from FIEs in China to MNCs' subsidiaries in regions with lower business risks. Under the current Chinese tariff regulations, materials and intermediate products imported by FIEs, which are used for production of goods oriented to the domestic market, are subject to an import tariff. This has the effect of curbing MNCs' transfer pricing activities.

In general, MNCs with export-oriented DFI have a higher motivation, capability and latitude to manipulate transfer pricing than MNCs whose DFI is oriented toward the Chinese domestic market. In the context of investors' countries of origin, investments from Hong Kong, Taiwan and Southeast Asian countries are largely labor-intensive and export-oriented. By comparison, investments from the U.S. and European countries are primarily capital/technology intensive and are oriented to the Chinese domestic market. The former is motivated by reducing the costs of production for exports by taking advantage of China's cheap and abundant labor resource, whereas the latter aims at China's huge domestic market. In the particular circumstances of China, although transfer pricing manipulation may exist in all types of FIEs, the frequency, scope and latitude of transfer pricing manipulated by MNCs based in Hong Kong and Taiwan would tend to be more prevalent than that of American and European MNCs.

8.4.3 Some Indirect Evidence

The study of transfer pricing involves an investigation of firms' pricing policies, market orientation, accounting systems and the internal financial transfers within a MNC system. As this information is treated in commercial confidence by firms, it is therefore unlikely to be obtained by researchers or others. As a result, most of the studies of MNCs transfer pricing either analyze theoretically the conditions under which MNCs perform transfer pricing, or present case studies using sampled data (e.g. Lecraw, 1985). In the case of China, there are more than 140,000 foreign-invested enterprises operating in various industries. A detailed investigation of these FIEs' pricing behavior at the level of the firm is impracticable, and beyond the scope of this chapter.

However, an investigation of the unit prices of China's exports to, and imports from, a country (or region) where most foreign investing companies are based is a feasible alternative. Trade between China and Hong Kong provides a suitable case for the study of unit price ratios of the same commodities.[5] This is because Hong Kong is the most important source of DFI in China and is also a major trading partner. The trade between China and Hong Kong is largely controlled by Hong Kong based MNCs and in recent years, FIEs have become major players in China-Hong Kong bilateral trade. In the coastal provinces, trade with Hong Kong is dominated by FIEs. Therefore, differences in the unit prices of the same commodities traded between China and Hong Kong are important indicators of transfer price manipulation by MNCs.

By comparing the unit prices of China's exports to Hong Kong which are reported to the Chinese Customs, with the unit prices of Hong Kong's imports reported to the Hong Kong Customs, we can measure the differences in the unit prices of the same commodities recorded by the two Customs authorities. Similarly, we can also measure the differences in the unit prices of China's imported commodities from Hong Kong, which are also reported to the two Customs authorities.

To study the unit prices of commodities traded between China and Hong Kong, a selection criteria for the traded commodities to be covered by this study has been adopted, viz: (i) the commodities are recorded by both physical measures (e.g. metric ton, number, length and height, etc.) and monetary values (i.e. total prices), so that the unit prices can be calculated; (ii) the physical measures of commodities reported in both the Chinese and Hong Kong trade statistics are close to each other so that meaningful comparisons of the unit prices of the commodities can be made. Commodities with no record of physical measures or where these measures are significantly different between the Chinese and Hong Kong statistics are excluded from this study. The data source is the *Commodity Trade Statistics* published by the United Nations in 1994.

Using this data source and the above selection criteria, the unit prices of 150 categories of exports reported in the Chinese Customs in 1994, as compared to the unit prices of the same categories of commodities reported in the Hong Kong Customs, have been calculated. The two sets of unit prices of the Chinese exports to Hong Kong (imports for Hong Kong) recorded respectively in the Chinese Customs and the Hong Kong Customs in 1994 and their ratios are presented in Appendix Table 8.1. In addition to the individual unit price ratio for each category of commodity, the weighted average unit price ratios for the 150 categories of commodities in 1994 are also calculated, using the value share of each commodity category in the total value of the listed exports as the weight.

As shown in Appendix Table 8.1, in 1994 the weighted average ratio of the unit price A (recorded by the Chinese Customs) to the unit price B (recorded by the Hong Kong Customs) is 0.826 (86.2%) for the 150 categories of exports, as shown in Appendix 9.1. This indicates that the prices of Chinese exports were lowered by an average 17.4 percent. This verifies that the prices of China's exports to Hong Kong are underinvoiced (underpriced) by Hong Kong traders.

Before we link this fact to possible transfer pricing manipulation by MNCs, two factors need to be taken into account. First, not all of China's exports are provided by FIEs. For example, FIEs' exports accounted for 32 percent of China's total exports in 1995 (see Table 8.1). A larger share of China's exports are from domestic firms. Thus, the unit price ratio of commodities traded between China and Hong Kong does not necessarily reflect fully the extent to which MNCs practice transfer price. However, most of the trade between China and Hong Kong is controlled by Hong Kong based trading companies and many of them have investments and subsidiaries in China. Accordingly, for the particular case of trade between China and Hong Kong, FIEs' share should be much larger than the average share of FIEs in the total exports or imports of China. For example, in Guangdong province where Hong Kong investment is concentrated, exports by FIEs accounted for 44 percent of the total provincial exports in 1995. FIEs' imports accounted for 56 percent of Guangdong's total imports in that year (see Table 8.1). Therefore, the unit price ratios of commodities traded between China and Hong Kong can be reasonably used as an indicator of transfer prices manipulated by MNCs based in Hong Kong.

Secondly, transport and insurance fees are components of the total prices of exports and imports. These fees may also influence the unit prices of commodities. Since the prices of exports are calculated using the f.o.b. (free on board) price formula which do not include transportation and insurance fees, they are normally lower than the prices of the same goods recorded as imports at the destination country. This is because the prices of imports are calculated using the c.i.f. price (cost, insurance and flight fees) formula. The import prices are in general higher than the prices of the same goods recorded as exports from the country of origin.

The difference between the two price measurements reflects the distance from the exporting country to the importing country, and insurance policies relative to trade and transportation. According to a recent estimation by Guangdong's Foreign Trade Bureau, transport fees account for around five percent of the commodities' prices (depending on the distance from Hong Kong) and the insurance fee accounts for only one percent of commodity prices. Therefore, in general the sum of transport and insurance fees are about six percent. Since this proportion is significantly smaller than the difference between the prices reported at the Chinese Customs and the prices reported at the Hong Kong's Customs, transfer prices of around 10 percent on average appear to exit in this bilateral trade which is dominated by Hong Kong invested FIEs.

In addition to exports, China's imports from Hong Kong are also subject to MNCs transfer pricing. This can be examined by comparison of the unit prices of China's imports (from Hong Kong) reported at the Chinese Customs with the unit prices of the same commodities reported at the Hong Kong Customs. The two sets of unit prices and their ratios of China's imports from Hong Kong in 1994, as recorded by the two Customs authorities, are presented in Appendix 8.2. As can be seen in the table, the unit prices of China's imports (from Hong Kong) reported at the Chinese Customs are generally higher than that reported at the Hong Kong Customs.

Since the price of imports is calculated by C.I.F. (viz. cost of commodities plus insurance and flight fees), the reported prices at the importing country's Customs should be slightly higher by the sum of insurance and transport fees than the prices reported (using F.O.B price) at exporting country's Customs. The weighted average ratio of the unit prices of the 66 categories of commodities reported at the two Customs Authorities in 1994 is 1.15 (115%). This ratio less the transport and insurance fees as a percentage of the total price (about 6 percent) provides a guide to overinvoicing of Chinese imports by MNCs in the China-Hong Kong bilateral trade of around 9 percent.

While these figures only relate to 1994 statistics, and may differ slightly from year to year, they point to Hong Kong trading companies, especially those investing in China, overcharging Chinese importers, including FIEs, by overinvoicing imports. As imports by Chinese domestic firms are subject to relatively high import duties, these firms do not have any special incentive to overinvoice their imports. Therefore, the overinvoicing of imports is primarily performed by MNCs which have either direct investment in China or trade relations with China. This case study suggests that transfer pricing manipulated by MNCs prevails in both exports and imports of China, although the extent may differ between regions and countries of origin of MNCs.

The effects of transfer pricing on the Chinese economy and social welfare are clearly negative. They reduces the taxation revenue for the Chinese government and lessen profits accruing to Chinese partners in joint ventures. As a result of transfer pricing, many FIEs report accounting losses to the taxation authority each year. For instance, 32.3 percent of FIEs in 1992 and 31.2 percent of FIEs in 1993, reported losses, which was an even higher rate than that of state-owned enterprises (23.3 percent and 28.8 percent respectively).[6] By 1995, 40 percent of FIEs in the industry sector reported an 'operating loss', which was still higher than that of state-owned industrial enterprises (33.8 percent).[7] Strangely, in spite of this upsurge in 'operating losses' reported by FIEs, many foreign investors have increased their investments in these 'loss-making' enterprises. These conflicting facts reveal that the 'operating losses' of some FIEs may have been artificially created by MNCs through transfer pricing.

The existence of transfer pricing is also evidenced by the fact that on average the pre-tax profit rate in FIEs is lower than that in state-owned enterprises. Chinese state-owned enterprises are characterized by low efficiency and low profitability, with many of them relying on fiscal subsidies from the government. It is therefore remarkable that the pre-tax profit rate in FIEs is even lower than that in such state-owned enterprises. For instance, the pre-tax profit rate (weighted average) for FIEs in the industry sector was 5.27 percent in 1992, compared to 11.38 percent in state-owned enterprises. In 1995, the pre-tax profit rate in FIEs was 7.8 percent, which was sill lower than 11 percent for state-owned enterprises (SSB, 1997)

The above evidence indicates that MNCs practice transfer pricing for exports and imports of FIEs at the expense of the Chinese government and local partners. This would have different impacts on the two parties. On the Chinese side, MNCs transfer pricing results in losses of taxation revenue and profit. On the part of foreign investing companies, the manipulation of transfer pricing serves their goals to maximize global profits and minimize their business costs and risks.

In terms of the economics of multinational corporations, transfer pricing by MNC's is a response to market imperfection caused by government interventions including tax, tariff and other risks. Transfer pricing activities tend to improve the overall efficiency of allocation of resources in the world (Chudson, 1985). In terms of income distribution, however, transfer pricing is harmful to the host country, especially for a developing host country which possesses a poorly equipped legal system and lacks professional expertise.

8.5 Conclusion and Policy Implications

This chapter has investigated the impact of DFI on the foreign trade of China. It has found that in general, DFI is trade-creating in China's case, and supports the Kojima hypothesis that foreign investments concentrated in the host country's comparatively advantaged industries tend to promote trade. As found from the empirical study, DFI has played an important role in Chinese export expansion, and has been a catalyst to economic growth in the past 18 years.

However, the increasing trade deficits of foreign-invested enterprises partly offset the trade surplus created by Chinese domestic sectors, and this partly negates the positive impact of FIEs export boom. Among the positive factors contributing to the persistent trade deficit of FIEs is the importation of capital goods tends which is necessary for domestic capital formation and economic growth. However, transfer pricing of FIEs' exports and imports has a negative impact on the Chinese economy, especially by reducing the government's taxation revenue and local partner's operating profits. This study indicates that under the current Chinese legal framework and economic environment, MNCs have a higher propensity to use transfer prices for FIEs' exports and imports. The evidence presented confirms the existence of transfer pricing in FIEs' foreign trade.

Some policy implications can be drawn upon the findings of this study. DFI, especially those investments in the host country's comparatively advantaged industries should be particularly encouraged. To reduce the trade deficit of FIEs, some polices such as input localization and the removal of tariff-free imports of intermediate inputs for FIEs, which can be produced domestically, can be considered. In addition, the government should further liberalize the economy and develop a well-defined legal system which governs business operating environment. As business environment improves, including reducing business risks and increasing operating predictability, transfer pricing by MNCs tends to diminish.

Notes

1   Calculated by using the 1981 exchange rate (1 US$ = 1.7051 *Renminbi*, the Chinese currency).
2   Calculated from Statistical Yearbook of China 1996.
3   *Economic Daily (Jingji Robao)*, May 17, 1995, p.2., which cited from SSB data.
4   The Chinese government recently modified the tariff policy on FIEs' imports. Under the new rule, the standard tariff will be applied to the imports of FIEs including capital goods and other products since 1997.
5   Although Hong Kong has become a part of China since July 1, 1997, the trade and Customs management in the mainland and Hong Kong are still independent from each other. Flows of commodities and capital between the two are still treated as 'international' trade and investment. The Customs statistics of commodity trade in the two sides are separate. Therefore, the unification of Hong Kong to the mainland is expected to have little impact on firms' pricing behavior for traded goods.
6   SSB: Annual Statistical Report of China's Industry 1992, 1993.
7   SSB (1997), *The Third National Industrial Census of People's Republic of China in 1995*.

9 Conclusions and Implications

9.1 Conclusions

This study has provided a systematic investigation of the impact of direct foreign investment on the economic development of China for the period 1979 to 1996. It has focused on the macroeconomic effects of DFI on the Chinese economy, such as promotion of domestic capital formation and economic growth, increasing exports and industrial production, creating new employment and reinforcing inter-regional economic disparity. In addition, the microeconomic impacts of DFI have been explored, including the entry modes of MNCs into Chinese market, the comparative efficiency of Chinese local firms and foreign-invested enterprises, and the evidence of transfer pricing manipulated by multinational corporations. The major findings of this study can be summarized as follows.

China has experienced an exceptionally high growth in DFI in the past eighteen years. The major investing countries (or regions) are Hong Kong, Taiwan, Japan and the United States. DFI has been concentrated in the Eastern (coastal) Region, with focus on labor-intensive manufacturing industries and the real estate sector. Each investing country has presented particular features in investment. Hong Kong and Taiwan investments have been concentrated in labor-intensive manufacturing industries which are largely oriented to exports. These enterprises are primarily located in the coastal region. By comparison, American and European investments have been primarily in capital and technology intensive industries, and target the Chinese domestic market, with enterprises being located in major cities.

In terms of the entry modes of foreign investments, this study has found that the cultural distance between investing countries and the host country negatively related to wholly foreign-owned enterprises or foreign majority share in joint ventures. A cultural proximity, as in the cases of Hong Kong and Taiwan investments in China, encourages full foreign ownership and foreign majority equity share in joint ventures. It has also shown that the technological nature and content of investment projects are positively associated with foreign ownership. A liberalized economic climate and favorable investment environment tend to encourage foreign capital involvement, especially wholly-owned subsidiaries.

There is little doubt that the impact of DFI on domestic capital formation, economic growth, export expansion, industrial production and employment in China has been positive. Furthermore, DFI has introduced a dynamic force into the Chinese economy through technology transfer and training, industrial linkage effects, more access to international markets, which have promoted market competition and productive efficiency. Therefore, foreign-invested enterprises (FIEs) have become a dynamic part of the Chinese economy and play a leading role in economic growth.

A comparative analysis of productive efficiency between FIEs and Chinese domestic firms indicates that FIEs generally have a higher capital-labor ratio, a larger incremental output-capital ratio, and greater average and marginal productivity of labor and capital. In addition, FIEs have a higher export propensity. Both FIEs and domestic firms share a similar production function. The estimation has shown that the elasticity of output with regard to change in labor and capital inputs was slightly larger in FIEs than in state-owned enterprises. This indicates that FIEs are generally more productive than domestic firms.

As a result of the dynamic role of FIEs, the overall allocative efficiency of economic resources in China and the productive efficiency in Chinese enterprises have been improved. This has also assisted economic transformation from a centrally-planed system to a market economy.

Notwithstanding these enormous benefits, this study has found that DFI has certain negative impact on the Chinese economy. One apparent negative impact is that DFI has contributed to increased inter-regional economic disparity within China. As a result of coast-oriented distribution supported by government policies, DFI has reinforced factors behind the existing inter-regional economic disparity, such as divergent capital formation, technology gap, and the differences in industrial structure and human capital. Another negative impact is transfer pricing by MNCs and resultant income losses accruing to the Chinese government and enterprises.

9.2 Policy Implications

Some policy implications can be drawn from the findings of this study for both China and other developing countries. First, DFI can play a positive role in economic development in a host country, provided that appropriate economic policies, especially DFI policies, can be formulated and implemented in the host country. The host government's capacity to maintain the political stability and sustainable economic growth is the prerequisite for DFI to positively affect the economy. In this regard, the Chinese experience in the past 18 years can be regarded as successful, although this country has still a long way to go to further economic reforms and to establish a well-defined legal system.

Second, the Chinese experience indicates that the formation of joint ventures is an effective way for a developing host country to use foreign investment. Since the joint venture mechanism integrates foreign investing firms with local firms in capital, technology and management, it promotes domestic capital formation and technology transfer. It also stimulates the market-based economic transformation including reforms of the ownership of state-owned enterprises.

To serve the goal of industrialization strategy, it is essential for the host government to formulate a suitable industry policy to guide the flow of DFI. The industrial distribution of DFI should be consistent with the industrial policy orientation and economic development goal of the host country. On the one hand, the host government should encourage DFI in high-technology industries which otherwise would develop slowly with domestic firms. To promote DFI in high-technology industries, some preferential policies are necessary, e.g. tax concessions for foreign-invested enterprises and more access to the domestic market for their products. In the Chinese case, DFI projects in industries such as automobile, computer and other advanced electronic products are primarily for the domestic market. As the domestic capacity for producing these high-technology products improves, imports can be substituted.

On the other hand, DFI in the industries in which the host country has comparative advantages should also be encouraged. For many developing countries, labor-intensive industries are the industries in which they possess comparative advantages, and therefore have strong competitiveness in international markets. DFI in these industries should be oriented to overseas markets, rather than targeting the host country's domestic markets. The host government may adopt legal and financial measures, such as export requirement and differential tax treatments, to encourage export-oriented DFI. Therefore, the promotion of export-oriented DFI in the industries of international competitiveness and encouragement of DFI in high-technology industries should be the DFI policy orientation for a developing host country.

The concurrent policy measures of export promotion and import-substitution accord with the industrialization strategy of a developing country.

In addition, to bring the backward linkage effects of DFI into full play, the host country should introduce DFI into the key industries that have high backward linkage indices and encourage foreign firms to use locally-made intermediate inputs in their production. As the local content of inputs used in foreign-invested enterprises rises, the backward linkage and multiplier effects of DFI on the domestic sectors will materialize and therefore strongly stimulate economic growth. To this end, the host country needs to introduce some policy measures such as preferential tax treatment for use of locally-made inputs.

Furthermore, the impact of DFI on inter-regional economic disparity is an important issue for host government policy. This is a critical matter for a developing country, especially for a country where inter-regional economic disparity has already existed. To eliminate the exacerbating impact of DFI on the inter-regional disparity, the host country should make efforts to introduce DFI into less-developed regions. Some special or favorable policies for DFI in the less-developed regions are required. These measures include a longer 'tax holiday' or tax reduction period, lower land use fees and lower charges for use of public utilities. Promotion of DFI should be regarded as an essential step to achieve the regional economic development strategy. In addition, fiscal support from the central government for major infrastructure construction in less-developed regions should be put on the government agenda. In the long run, the balanced development of all regions should become the policy priority for national economic and social development.

Finally, transfer pricing manipulated by foreign investors undermines the interests of the host country. It is necessary for the host country to further reform economic system, liberalize business operating environment and establish a well-defined legal framework so as to reduce business risks. This tends to reduce MNCs' motivation to use transfer prices. On the part of the host country's interest, the host government may use legal and administrative instruments to restrict transfer pricing. In addition, international cooperation between investing countries and the host country in taxation, pricing, accounting, auditing and tariff management should be developed to deal with this matter.

Appendixes

Appendix to Chapter 3

Table 3.1:   Registered Foreign-Invested Enterprises in China by Entry Mode at the End of 1993 (US$ millions)

	EJVs		CJVs		WFOEs		Total	
	Value	%	Value	%	Value	%	No.	Value
Agriculture[1]	994	40.4	869	35.3	600	24.4	4246	2462
Industry[2]	46422	55.2	15051	17.9	22553	26.8	124606	84026
Geological Exploration	19	94.9	0.2	2.2	0.7	3.8	47	20
Construction	1410	42.9	1057	32.2	817	24.9	4603	3284
Transport, Post & Telecommunication	1810	64.0	966	34.2	50	1.8	1918	2826
Commerce, Catering & Storing	3711	49.5	2079	27.7	1706	22.8	8742	7496
Real estate, hotel	18679	40.2	14497	31.2	13254	28.5	19384	46431
Finance & insurance	113	44.8	0	0.0	139	55.2	31	252
Health care, sports & social welfare	146	27.0	322	60.0	71	13.3	357	539
Education, culture	548	46.1	351	29.4	291	24.5	1609	1190
Scientific services	222	50.9	39	8.9	175	40.1	878	435
Other	616	50.5	226	18.5	380	31.1	1086	1222
All Sectors	74690	49.7	35456	23.6	40036	26.7	167507	150182

Notes: EJV stands for equity joint venture, CJV for contractual joint venture, and WFOE for wholly foreign-owned enterprise. 1.This sector includes agriculture, forestry, fishing and husbandry. 2. Industry sector includes manufacturing and mining. 3. The 'No.' refers to number of investment projects.

Source: SSB *Foreign Economic Statistical Yearbook*, 1994, pp. 311-314.

Table 3.2:   Registered Foreign-Invested Enterprises in China in 1992 by Entry Mode and Industry

(in Millions of US Dollars)

Industries	EJVs[1]				CJVs				WFOEs				Total	
	No.	%	Value	%	No.	%	Value	%	No.	%	Value	%	No.	Value
Food[2]	2141	75.0	1331	62.9	316	11.1	306	14.5	362	12.8	479	22.7	2819	2116
Textile, sewing	4625	69.1	2355	53.9	882	13.2	751	17.2	1190	17.8	1236	28.4	6697	4342
Light manufacture	3244	65.1	1661	52.8	704	14.1	560	17.8	1036	20.8	923	29.4	4984	3144
Chemicals[3]	3503	78.6	2182	65.6	394	8.8	385	11.5	592	13.2	789	23.5	4489	3356
Pharmaceutical	554	89.2	402	85.6	43	6.5	24	0.7	67	10.1	68	14.4	664	494
Machinery and electronics	4680	70.1	2830	52.6	662	9.6	821	15.3	1326	19.8	1725	32.1	6668	5376
Other	3657	89.3	1087	70.1	121	3.0	124	8.0	317	7.7	341	22.0	4095	1551
Total Manufacture	30471	73.6	11955	58.6	3122	10.2	2971	14.6	4925	16.2	5483	26.9	30471	20409

Notes: 1. EJV stands for equity joint venture, CJV for contractual joint venture, and WFOE for wholly foreign-owned enterprise.
2. This industry includes food, beverage, tobacco and forage processing. 3. This industry includes chemical materials and products, rubber and plastic products. The 'No'. refers to the number of investment projects.

Source: MOFTEC 1993.

Appendix to Chapter 4

Table 4.1: FIE's Contribution to China's Industry Growth (100 million *yuan*)

Indicators	1988			1990			1993			1995		
	Total	FIE	%	Total	FIE	%	Total	FIE	%	Total	FIE	%
Gross Output Value	17458	156	0.9	19637	378	2.0	34048	3460	10.2	46883	9082	7.9
Value Added	4301	36	0.8	5093	97	1.9	12843	1081	8.4	15446	2282	14.8
Original Value of Fixed Capital	10641	63	0.6	14390	221	1.5	25818	1686	6.5	44988	4904	10.9
Net Value of Fixed Capital	7420	56	0.8	10139	187	1.8	18428	1363	7.4	32287	3883	12.0
Pre-tax Profit	2289	19	0.8	2235	36	1.6	3924	327	8.3	5050	701	13.9
Net Profit	1190	13	1.1	560	22	3.9	964	169	17.5	1114	313	28.1
Tax paid	489	2	0.4	414	5.8	1.4	2321	158	6.8	4034	398	9.9
No. of Enterprises (hundred)	4209	11	0.3	4171	25	0.6	4492	201	0.5	5104	443	8.7
Employees (thousand)	75180	239	0.3	7481	51	0.7	8299	326	0.9	101695	8031	7.9

Note: The figures for 1988-1990 refer to all enterprises with independent accounting system, the figures for all other years refer to the enterprises with independent accounting system at town level and above. The gross output value in this table is at 1990 constant price. Value for other indicators are at current prices.

Source: SSB, *Zhongguo Gongye Tongji Nianbao (The Annual Statistical Report of China's Industry)* 1988-1993, and *The Data of The Third National Industrial Census of The People's Republic of China in 1995.*

Appendix to Chapter 6

Table 6.1: Open Inverse Matrix Column Output Linkage Indices in China in 1992

Sector	Column Total	Column Mean	Standard Deviation	Coefficient Variation	Backward Linkage	Backward Spread
Agriculture	1.8345	0.0556	0.2086	3.7521	0.6859	1.4723
Coal mining	2.4967	0.0757	0.1829	2.4177	0.9335	0.9487
Oil. Gas exploring	2.0196	0.0612	0.1781	2.9106	0.7551	1.1421
Metal mining	2.5865	0.0784	0.2004	2.5566	0.9671	1.0032
Non-metal mining	2.4692	0.0748	0.1826	2.4406	0.9233	0.9577
Food processing	2.6036	0.0789	0.2187	2.7724	0.9735	1.0879
Textile	3.1623	0.0958	0.2840	2.9638	1.1824	1.1630
Apparel, shoes	3.2253	0.0977	0.2114	2.1628	1.2060	0.8487
Wood Products	3.0665	0.0929	0.2087	2.2454	1.1466	0.8811
Paper, printing	2.9655	0.0899	0.2194	2.4416	1.1088	0.9581
Power, water	2.2929	0.0695	0.1818	2.6169	0.8574	1.0269
Petroleum	2.5372	0.0769	0.1998	2.5992	0.9487	1.0199
Coal Products	2.9355	0.0890	0.1845	2.0744	1.0976	0.8140
Chemicals	2.9187	0.0884	0.2622	2.9646	1.0913	1.1633
Build materials	2.7294	0.0827	0.1963	2.3739	1.0205	0.9315
Metal smelting	2.9265	0.0887	0.2564	2.8914	1.0942	1.1346
Metal Products	3.1432	0.0952	0.2133	2.2394	1.1753	0.8787
Machinery	3.0296	0.0918	0.2335	2.5435	1.1328	0.9980
Transport Equip	3.1138	0.0944	0.2359	2.4999	1.1643	0.9810
Electrical	3.1155	0.0944	0.2142	2.2691	1.1649	0.8904
Electronics	3.1750	0.0962	0.2619	2.7222	1.1872	1.0682
Instruments	2.8605	0.0867	0.1943	2.2411	1.0696	0.8794
Machine Repairing	2.9942	0.0907	0.1830	2.0167	1.1196	0.7914
Oth manufacturing	3.1097	0.0942	0.2066	2.1919	1.1628	0.8601
Construction	2.9415	0.0891	0.1804	2.0235	1.0999	0.7940
Transport, post	2.1717	0.0658	0.1795	2.7272	0.8120	1.0701
Commerce	2.3264	0.0705	0.1915	2.7159	0.8699	1.0657
Catering	2.4004	0.0727	0.1864	2.5632	0.8976	1.0058
Passenger transpt	2.2297	0.0676	0.1744	2.5816	0.8337	1.0130
Pub utilities	2.0889	0.0633	0.1809	2.8585	0.7810	1.1216
Culture, Education	2.2413	0.0679	0.1779	2.6199	0.8380	1.0280
Finance	2.1953	0.0665	0.1804	2.7120	0.8208	1.0642
Administration	2.3499	0.0712	0.1701	2.3890	0.8786	0.9374
Total	88.256	2.6744	6.7402	84.099	33.000	33.000
Average	2.6744	0.0810	0.2042	2.5484	1.0000	1.0000

Backward Linkage = Column Mean / Average Column Mean.
Backward Spread = Coefficient Variation / Average Coefficient Variation.

Table 6.2: Total Output Multipliers in China in 1992

Sector	Initial	First	Industrial	Total I	Consumption	Total II
Agriculture	1.000	0.356	0.479	1.834	0.179	2.013
Coal mining	1.000	0.561	0.935	2.497	0.637	3.133
Oil, gas exploring	1.000	0.378	0.641	2.020	0.734	2.753
Metal mining	1.000	0.607	0.980	2.586	0.572	3.158
Non-metal mining	1.000	0.559	0.911	2.469	0.472	2.941
Food & beverage	1.000	0.743	0.861	2.604	0.289	2.892
Textile	1.000	0.794	1.368	3.162	0.383	3.545
Clothing, shoes	1.000	0.788	1.437	3.225	0.362	3.587
Wood products	1.000	0.747	1.320	3.067	0.417	3.484
Paper, printing	1.000	0.730	1.235	2.965	0.388	3.353
Power, steam	1.000	0.512	0.781	2.293	0.755	3.048
Oil processing	1.000	0.718	0.819	2.537	0.601	3.138
Coal products	1.000	0.775	1.161	2.936	0.667	3.603
Chemicals	1.000	0.721	1.197	2.919	0.461	3.380
Building materials	1.000	0.653	1.076	2.729	0.452	3.182
Metal smelting	1.000	0.715	1.211	2.926	0.490	3.417
Metal products	1.000	0.760	1.383	3.143	0.427	3.570
Machinery	1.000	0.717	1.312	3.030	0.420	3.449
Transport vehicle	1.000	0.733	1.381	3.114	0.396	3.510
Electrical	1.000	0.746	1.369	3.115	0.406	3.522
Electronics	1.000	0.750	1.425	3.175	0.422	3.597
Instruments	1.000	0.661	1.199	2.860	0.392	3.252
Machine repairing	1.000	0.697	1.297	2.994	0.464	3.458
Other manufactures	1.000	0.762	1.347	3.110	0.408	3.517
Construction	1.000	0.704	1.237	2.941	0.377	3.318
Transport, post	1.000	0.435	0.737	2.172	0.593	2.764
Commerce	1.000	0.533	0.793	2.326	0.357	2.683
Catering	1.000	0.598	0.803	2.400	0.214	2.614
Passenger transport	1.000	0.455	0.775	2.230	0.543	2.773
Public utilities	1.000	0.415	0.674	2.089	0.748	2.837
Culture, education	1.000	0.465	0.776	2.241	0.449	2.691
Finance	1.000	0.478	0.717	2.195	0.351	2.546
Administration	1.000	0.522	0.828	2.350	0.443	2.793

Appendix to Chapter 7

Table 7.1: Real GDP and Growth Rates in Different Regions of China
(at the 1980 constant price)

Regions	Real GDP[1] (100 million *yuan*)					Population (million people)		
	1980	1985	1993	1995	1980-95 growth %	1980	1995	1980-95 growth %
The East	**2187.5**	**3693.6**	**8632.9**	**11444.7**	**11.7**	**362.29**	**450.56**	**1.46**
Guangdong	245.7	436.1	1376.4	1883.5	14.5	52.27	68.68	1.84
Fujian	85.9	147.9	390.0	546.8	13.1	25.18	32.37	1.69
Jiangsu	319.8	568.3	1414.5	1901.7	12.6	59.38	70.66	1.17
Zhejiang	179.7	357.5	888.8	1244.7	13.8	38.27	43.19	0.81
Shanghai	311.9	466.8	897.2	1170.1	9.2	11.46	14.15	1.42
Shandong	301.2	509.3	1266.9	1682.6	12.2	72.96	87.05	1.18
Hebei	219.2	355.4	763.3	998.9	10.6	51.68	64.37	1.47
Beijing	139.1	216.0	438.3	559.1	9.7	9.04	12.51	2.19
Tianjin	103.5	161.5	276.7	363.4	8.7	7.49	9.42	1.54
Liaoning	281.5	432.5	805.6	959.4	8.5	34.56	40.92	1.13
Hainan	n/a	42.3	115.2	134.5	n/a	n/a	7.24	n/a
The West[2]	**789.4**	**1284.5**	**2474.8**	**2772.3**	**9.4**	**245.42**	**299.54**	**1.34**
Shaanxi	95.2	159.8	309.2	365.7	9.4	28.31	35.14	1.45
Sichuan	322.0	507.8	945.3	1155.3	8.9	99.20	113.25	0.87
Qinghai	18.3	28.1	44.7	52.2	7.2	3.77	4.81	1.64
Yunnan	84.3	147.2	300.9	373.4	10.4	31.73	39.90	1.54
Guizhou	60.3	107.8	196.5	229.2	9.3	27.77	35.08	1.57
Ningxia	15.1	25.7	46.2	54.5	8.9	3.69	4.38	1.15
Gansu	73.9	106.2	235.1	285.2	9.4	19.18	25.13	1.82
Inner Mongolia	67.9	105.0	187.0	211.37	7.9	18.94	22.84	1.26
Xinjiang	53.0	96.9	209.9	256.8	11.1	12.83	16.61	1.74

Notes: 1. The deflators used in calculating the real GDP for each province are their provincial deflators for each year. 2. Since the data for Tibet are incomplete, it is excluded in the figures of the West Region.

Source: The statistical yearbooks of all the province listed above for period from 1986-95, and Statistical Yearbook of China 1995 and 1996.

Appendix to Chapter 8

Table 8.1: Unit Price Ratios of China's Exports to Hong Kong's Imports in 1994

SITC	Unit	China's Exports to Hong Kong			Hong Kong's Imports from China			Unit Price Ratio		
		Volume	Value ($1000)	Unit Price A	Volume	Value ($1000)	Unit Price B	Price Ratio A/B	Weight %	Weighted Price Ratio
011	W	11433	15201	1.330	13272	19506	1.470	0.905	0.097	0.088
012	W	83901	95799	1.142	62444	98066	1.570	0.727	0.612	0.445
016	W	6077	18339	3.018	5953	16822	2.826	1.068	0.117	0.125
017	W	38607	65337	1.692	41513	67457	1.625	1.041	0.418	0.435
017.2	W	6883	12448	1.809	6460	12252	1.897	0.954	0.080	0.076
017.4	W	3269	8090	2.475	3794	9738	2.567	0.964	0.052	0.050
017.5	W	21598	33477	1.550	26119	37606	1.440	1.077	0.214	0.230
022	W	12782	7550	0.591	26192	19202	0.733	0.806	0.048	0.039
025	W	30909	22837	0.739	30023	25704	0.856	0.863	0.146	0.126
034.2	W	12496	31112	2.490	6682	13902	2.081	1.197	0.199	0.238
035	W	3532	25570	7.240	2954	36046	12.202	0.593	0.163	0.097
036	W	37563	151690	4.038	31219	113716	3.643	1.109	0.970	1.075
037	W	8869	28083	3.166	6113	17945	2.936	1.079	0.180	0.194
042	W	38589	13125	0.340	38683	17222	0.445	0.764	0.084	0.064
044	W	31141	3321	0.107	5993	807	0.135	0.792	0.021	0.017
046	W	8259	1932	0.234	8061	2168	0.269	0.870	0.012	0.011
047	W	7876	2231	0.283	8678	2532	0.292	0.971	0.014	0.014
048	W	51669	42050	0.814	46519	42281	0.909	0.895	0.269	0.241
054	W	492045	174435	0.355	304255	160331	0.527	0.673	1.115	0.750
057	W	121196	97112	0.801	162877	133550	0.820	0.977	0.621	0.607
058	W	20995	28788	1.371	33858	37772	1.116	1.229	0.184	0.226
061	W	75899	26044	0.343	47717	19073	0.400	0.858	0.167	0.143
062	W	21608	29706	1.375	15472	24895	1.609	0.854	0.190	0.162
071	W	1097	3733	3.403	1078	3675	3.409	0.998	0.024	0.024
074	W	19213	39520	2.057	20871	43781	2.098	0.981	0.253	0.248
081	W	212020	40298	0.190	189111	34496	0.182	1.042	0.258	0.268
111	MW	698519	184648	0.264	683152	188191	0.275	0.960	1.181	1.133
112	W	26740	23168	0.866	25395	32509	1.280	0.677	0.148	0.100
121	W	4306	9085	2.110	4954	12476	2.518	0.838	0.058	0.049
122.2	W	14033	229011	16.32	23742	347352	14.630	1.115	1.464	1.633
222	W	26983	15556	0.577	28962	19074	0.659	0.875	0.099	0.087
223	W	26983	15556	0.577	28962	19074	0.659	0.875	0.099	0.087
261	W	5462	95342	17.46	5648	104280	18.463	0.945	0.610	0.576
263	W	11927	13882	1.164	13680	11677	0.854	1.364	0.089	0.121
266	W	2069	3322	1.606	3696	5302	1.435	1.119	0.021	0.024
268	W	4923	72450	14.72	5611	84808	15.115	0.974	0.463	0.451

269	W	4548	2828	0.622	5504	2525	0.459	1.355	0.018	0.025
273	MW	12608	37836	3.001	10929	50154	4.589	0.654	0.242	0.158
278	MW	1252	27127	21.67	884	22302	25.229	0.859	0.173	0.149
291	W	7491	51037	6.813	10341	67135	6.492	1.049	0.326	0.342
292.3	W	47111	10202	0.217	46531	11365	0.244	0.887	0.065	0.058
321	MW	1600	44422	27.76	1411	46668	33.074	0.839	0.284	0.238
334	W	447033	75457	0.169	401093	66877	0.167	1.012	0.482	0.488
421	W	248418	190628	0.767	105190	88014	0.837	0.917	1.219	1.118
422	W	383294	242445	0.633	103142	60183	0.583	1.084	1.550	1.680
431	W	4828	1394	0.289	5569	3963	0.712	0.406	0.009	0.004
513	W	26752	31042	1.160	37604	35923	0.955	1.215	0.198	0.241
515	W	7332	65151	8.886	9801	76947	7.851	1.132	0.417	0.471
523	W	25147	6464	0.257	19895	6152	0.309	0.831	0.041	0.034
524.3	W	7502	18192	2.425	6751	17381	2.575	0.942	0.116	0.110
531	W	19197	74079	3.859	17585	69347	3.944	0.979	0.474	0.463
533.2	W	4371	10962	2.508	3195	7718	2.416	1.038	0.070	0.073
533.4	W	39336	59507	1.513	36252	57443	1.585	0.955	0.380	0.363
541.1	MW	4562	108557	23.68	4220	113491	26.894	0.885	0.694	0.614
553.4	W	2714	5641	2.078	2766	4780	1.728	1.203	0.036	0.043
554.1	W	6062	7505	1.238	8613	11600	1.347	0.919	0.048	0.044
554.2	W	30130	17988	0.597	26865	14345	0.534	1.118	0.115	0.129
562	W	12195	1803	0.148	14780	2855	0.193	0.765	0.012	0.009
571	W	7155	4856	0.679	11392	8052	0.707	0.960	0.031	0.030
572	W	48165	38108	0.791	51870	49391	0.952	0.831	0.244	0.202
573	W	61984	47160	0.761	59374	51848	0.873	0.871	0.302	0.263
574	W	22660	24080	1.063	17617	20773	1.179	0.901	0.154	0.139
575	W	11477	11495	1.002	18751	22053	1.176	0.852	0.073	0.063
581	W	14713	14593	0.992	13245	15545	1.174	0.845	0.093	0.079
592.1	W	13228	4075	0.308	11438	3054	0.267	1.154	0.026	0.030
613	W	559	25834	46.22	999	57425	57.482	0.804	0.165	0.133
621	W	5607	8234	1.469	2206	6576	2.981	0.493	0.053	0.026
625.2	N	94872	5024	0.053	84070	5417	0.064	0.822	0.032	0.026
625.4	N	1983	2143	1.081	7692	10871	1.413	0.765	0.014	0.010
641	W	103504	58441	0.565	96550	67503	0.699	0.808	0.374	0.302
651	W	300322	1016821	3.386	288653	1103213	3.822	0.886	6.501	5.759
655	W	148076	525101	3.546	160175	576491	3.599	0.985	3.357	3.308
656	W	14418	106034	7.354	14832	87446	5.896	1.247	0.678	0.846
657.2	W	2924	6898	2.359	5088	23972	4.711	0.501	0.044	0.022
657.3	W	16700	61565	3.687	15089	45754	3.032	1.216	0.394	0.479
657.5	W	21063	38097	1.809	14749	37043	2.512	0.720	0.244	0.175
658.1	W	62942	13279	0.211	31199	9707	0.311	0.678	0.085	0.058
658.2	MW	2253	12508	5.552	10020	69367	6.923	0.802	0.080	0.064
658.4	MN	401521	215083	0.536	638924	353610	0.553	0.968	1.375	1.331
658.5	MN	11519	20506	1.780	11600	21444	1.849	0.963	0.131	0.126
661	W	748935	74335	0.099	601650	52138	0.087	1.145	0.475	0.544
671	W	42892	19766	0.461	46082	20638	0.448	1.029	0.126	0.130
673	W	68997	26398	0.383	99977	36323	0.363	1.053	0.169	0.178
674	W	32782	19093	0.582	19604	11825	0.603	0.966	0.122	0.118

676	W	488695	132018	0.270	202014	83052	0.411	0.657	0.844	0.555
678	W	37299	19786	0.530	38610	23296	0.603	0.879	0.126	0.111
679	W	76874	47897	0.623	64643	42615	0.659	0.945	0.306	0.289
682	W	52898	149298	2.822	47360	145350	3.069	0.920	0.955	0.878
683	W	6235	35632	5.715	4887	27656	5.659	1.010	0.228	0.230
684	W	50597	84404	1.668	45995	82940	1.803	0.925	0.540	0.499
685	W	27820	18186	0.654	5567	4361	0.783	0.834	0.116	0.097
686	W	26857	22797	0.849	27865	26289	0.943	0.900	0.146	0.131
687	W	14221	61940	4.356	15772	69659	4.417	0.986	0.396	0.391
689	W	14450	52838	3.657	16708	59896	3.585	1.020	0.338	0.345
693	W	18223	20389	1.119	11146	11981	1.075	1.041	0.130	0.136
694	W	37557	38969	1.038	37277	42786	1.148	0.904	0.249	0.225
713.8	N	24453	5508	0.225	24487	8133	0.332	0.678	0.035	0.024
716.1	MN	694735	197061	0.284	968080	478563	0.494	0.574	1.260	0.723
716.3	N	5425	26246	4.838	7852	54218	6.905	0.701	0.168	0.118
724.7	N	12948	4993	0.386	23182	16633	0.717	0.537	0.032	0.017
725.1	N	376	1641	4.364	528	3225	6.108	0.715	0.010	0.007
731	N	49370	19912	0.403	13776	97676	7.090	0.057	0.127	0.007
733	N	8050	10687	1.328	8426	12348	1.465	0.906	0.068	0.062
743.4	MN	16498	156930	9.512	47718	453356	9.501	1.001	1.003	1.005
743.6	MN	1188	12006	10.11	1644	17809	10.833	0.933	0.019	0.018
744.2	N	82909	3042	0.037	95209	4966	0.052	0.703	0.019	0.014
744.3	N	69	6809	98.68	141	15071	106.89	0.923	0.044	0.040
746.1	MN	55635	29767	0.535	46792	28466	0.608	0.879	0.190	0.167
747.8	MN	9397	12456	1.326	10435	16323	1.564	0.847	0.080	0.067
751	MN	99568	171562	1.723	208136	534459	2.568	0.671	1.097	0.736
761	MN	1728	106949	61.89	8028	742105	92.440	0.670	0.684	0.458
762	MN	58953	445863	7.563	206234	2686099	13.025	0.581	2.851	1.655
763	MN	11222	141444	12.60	44191	731736	16.558	0.761	0.904	0.688
764.2	MN	675029	219997	0.326	679374	283729	0.418	0.780	1.407	1.098
771.2	YN	694	133699	192.6	1672	257026	153.72	1.253	0.855	1.071
775.1	N	93684	4701	0.050	39867	3152	0.079	0.635	0.030	0.019
775.2	N	71912	6718	0.093	66925	6798	0.102	0.920	0.043	0.040
776.1	N	928101	20126	0.022	752028	29499	0.039	0.553	0.129	0.071
776.4	YN	11081	133267	12.03	7036	105341	14.972	0.803	0.852	0.684
781	N	82	950	11.58	3519	70877	20.141	0.575	0.006	0.003
782	N	1307	14174	10.85	1796	28158	15.678	0.692	0.091	0.063
785.1	N	7494	3203	0.427	24246	34416	1.419	0.301	0.020	0.006
785.2	MN	813	21751	26.75	6778	253773	37.441	0.715	0.139	0.099
793.2	N	472	198447	420.4	82	47953	584.79	0.719	1.269	0.912
831.1	MN	70686	102604	1.452	296675	988904	3.333	0.435	0.656	0.286
831.2	MN	238547	377633	1.583	35914	256901	7.153	0.221	2.414	0.534
841	MN	659411	2392108	3.628	511454	1861291	3.639	0.997	15.294	15.245
842	MN	658637	2331004	3.539	488586	2686596	5.499	0.644	14.903	9.592
843	MN	279739	226896	0.811	518705	491507	0.948	0.856	1.451	1.242
844	MN	467585	385153	0.824	970826	785565	0.809	1.018	2.462	2.507
845.2	MN	19793	41664	2.105	49132	151882	3.091	0.681	0.266	0.181
845.3	MN	245637	747497	3.043	316180	1828491	5.783	0.526	4.779	2.515

845.4	MN	403606	421560	1.044	487259	546443	1.121	0.931	2.695	2.510
845.6	MN	15484	12724	0.822	80344	183367	2.282	0.360	0.081	0.029
845.8	MN	30733	127054	4.134	128186	620093	4.837	0.855	0.812	0.694
845.9	MN	51523	101551	1.971	80344	183367	2.282	0.864	0.649	0.561
851.3	MP	114381	132541	1.159	812202	2295017	2.826	0.410	0.847	0.347
851.4	MP	42415	193114	4.553	272277	2388965	8.774	0.519	1.235	0.641
851.5	MP	66105	69259	1.048	289524	402953	1.392	0.753	0.443	0.333
898.1	N	123240	3297	0.027	313155	6548	0.021	1.279	0.021	0.027
898.2	N	6273	11864	1.891	10684	21455	2.008	0.942	0.076	0.071
All listed above			1564122			2878106			100.0	82.625

Notes: N: number; M: 1000 times unit shown; W: weight, metric ton; A: area, 1000 square meters; L: length, 1000 meters; H: 1000 kwh; P: number of pairs; K: kilogram; Y: 1,000,000 times unit shown.

Source: United Nations Statistical Office: Commodity Trade Statistics 1994, China, Rev. 3; and Hong Kong, Rev. 3.

Table 8.2: Unit Prices of China's Imports from Hong Kong in 1994

SITC	Unit	China's Imports from Hong Kong			Hong Kong's Exports to China			Unit Price Ratio		
		Volume	Value ($1000)	Unit Price A	Volum	Value ($1000)	Unit Price B	Price Ratio A/B	Weight %	Weighted Price Ratio
036	W	1168	3415	2.924	355	657	1.851	1.580	0.173	0.273
046	W	6158	2250	0.365	5244	2014	0.384	0.951	0.114	0.108
048	W	5681	8155	1.435	4671	9752	2.088	0.688	0.413	0.284
062	W	6676	19807	2.967	2903	8706	2.999	0.989	1.002	0.992
081	W	28903	11648	0.403	6445	1212	0.188	2.143	0.589	1.263
098.9	W	10122	19768	1.953	6832	14795	2.166	0.902	1.000	0.902
111	W	27841	11377	0.409	25635	10480	0.409	1.000	0.576	0.575
112	W	1658	2099	1.266	10363	7350	0.709	1.785	0.106	0.190
251	W	316425	32719	0.103	295446	24917	0.084	1.226	1.656	2.030
263	W	10066	2981	0.296	7205	2406	0.334	0.887	0.151	0.134
266	W	30256	42053	1.390	1440	1711	1.188	1.170	2.128	2.489
269	W	5488	3119	0.568	2709	1456	0.537	1.057	0.158	0.167
282	W	75018	9832	0.131	200574	24689	0.123	1.065	0.497	0.530
288	W	139277	29076	0.209	85331	22965	0.269	0.776	1.471	1.141
351	MH	1719	102572	59.67	1856	107674	58.01	1.029	5.190	5.338
411	W	17742	4861	0.274	16846	5446	0.323	0.848	0.246	0.208
421	W	27110	15777	0.582	5303	3108	0.586	0.993	0.798	0.793
531	W	10458	63524	6.074	7713	40567	5.260	1.155	3.214	3.712
532	W	1286	2416	1.879	996	882	0.886	2.122	0.122	0.259
533.1	W	7360	16992	2.309	7204	17097	2.373	0.973	0.860	0.836
553.4	W	19844	38607	1.946	731	1555	2.127	0.915	1.954	1.787
554.1	W	1731	2422	1.399	889	1385	1.558	0.898	0.123	0.110
571	W	25604	18466	0.721	1546	1221	0.790	0.913	0.934	0.853
572	W	89229	81445	0.913	384150	308675	0.804	1.136	4.121	4.681
573	W	33930	28700	0.846	73929	65958	0.892	0.948	1.452	1.377
581	W	4167	8110	1.946	3035	4745	1.563	1.245	0.410	0.511
582.9	W	25137	42953	1.709	2922	4688	1.604	1.065	2.173	2.315
592	W	21745	22748	1.046	20729	15407	0.743	1.407	1.151	1.620
621	W	4403	8724	1.981	1339	2851	2.129	0.931	0.441	0.411
641	W	330121	146590	0.444	399725	168590	0.422	1.053	7.417	7.809
655	W	126671	278317	2.197	117302	248083	2.115	1.039	14.08	14.631
657.2	W	7150	25169	3.520	5498	17196	3.128	1.125	1.274	1.433
661	W	423584	35054	0.083	443335	29673	0.067	1.236	1.774	2.193
674	W	17764	10904	0.614	4641	2616	0.564	1.089	0.552	0.601
676	W	115424	42130	0.365	72132	22779	0.316	1.156	2.132	2.464
678	W	9355	6638	0.710	15259	10247	0.672	1.057	0.336	0.355
682	W	35948	89175	2.481	31564	75784	2.401	1.033	4.512	4.662
684	W	29299	43361	1.480	23401	31998	1.367	1.082	2.194	2.375
686	W	6967	7180	1.031	10912	10552	0.967	1.066	0.363	0.387
693	W	3871	6300	1.627	3224	4002	1.241	1.311	0.319	0.418
694	W	9984	21385	2.142	12009	21675	1.805	1.187	1.082	1.284

695	MN	2968	14236	4.796	7404	11996	1.620	2.960	0.720	2.133
696.6	MN	1206	4363	3.618	3613	6634	1.836	1.970	0.221	0.435
716.1	MN	146755	43996	0.300	77296	20518	0.265	1.129	2.226	2.514
716.3	N	597852	7709	0.013	184345	1762	0.010	1.349	0.390	0.526
724.7	N	8195	61315	7.482	6792	36618	5.391	1.388	3.103	4.306
726.6	N	5531	25523	4.615	2047	8968	4.381	1.053	1.291	1.360
731	N	17438	53919	3.092	12340	27688	2.244	1.378	2.728	3.760
733	N	10045	42205	4.202	6345	21017	3.312	1.268	2.136	2.709
743.1	N	127447	17666	0.139	44476	5603	0.126	1.100	0.894	0.984
743.4	N	400072	6228	0.016	314687	5105	0.016	0.960	0.315	0.302
743.6	N	83358	38355	0.460	183872	57240	0.311	1.478	1.941	2.869
746.1	MN	9716	7533	0.775	4435	2337	0.527	1.471	0.381	0.561
751	MN	4049	12434	3.071	5508	7887	1.432	2.145	0.629	1.349
752	N	394389	65435	0.166	298674	28202	0.094	1.757	3.311	5.818
761	N	21047	6676	0.317	28401	3804	0.134	2.368	0.338	0.800
762	MN	2232	7061	3.164	2034	6028	2.964	1.067	0.357	0.381
764.2	MN	176648	36327	0.206	23147	7329	0.317	0.649	1.838	1.194
771.1	MN	81827	29951	0.366	38332	10907	0.285	1.286	1.516	1.950
776.1	N	439884	23520	0.053	41697	2999	0.072	0.743	1.190	0.885
776.4	YN	1017	22478	22.10	2067	44431	21.50	1.028	1.137	1.169
821.5	N	196727	16333	0.083	82964	7000	0.084	0.984	0.826	0.813
831.2	MN	4570	5674	1.242	385	481	1.249	0.994	0.287	0.285
843	MN	1583	5552	3.507	1706	5105	2.992	1.172	0.281	0.329
845.3	MN	28932	43947	1.519	34514	50018	1.449	1.048	2.224	2.331
851.4	MP	1930	9031	4.679	1627	5908	3.631	1.289	0.457	0.589
All listed above					1976286				100.0	114.853

Notes: N: number; M: 1000 times unit shown; W: weight, metric ton; A: area, 1000 square meters; L: length, 1000 meters; H: 1000 kwh; P: number of pairs; K: kilogram; Y: 1,000,000 times unit shown.

Source: United Nations Statistical Office: Commodity Trade Statistics 1994, China, Rev. 3; and Hong Kong, Rev. 3.

Bibliography

Ahiakpor, James C.W. (1990), *Multinationals and Economic Development: An Integration of Competing Theories*, London and New York: Routledge Press.

Allen, Douglas W and Lueck, Dean (1993), Transaction costs and the design of cropshare contracts, The Rand Journal of Economics, Vol.24, No.1, Spring, pp.78-100.

Anderson, Erin and Gatignon, Hubert (1986), Modes of foreign entry: a transaction cost analysis and propositions, *Journal of International Business Studies*, Vol.17, No.3, pp.1-26.

Areskoug, Kaj (1976): Private foreign investment and capital formation in developing countries. *Economic Development and Cultural Change*, Vol. 24, No. 3, April, pp. 539-547.

Arndt, H. W. (1974), Professor Kojima on the macroeconomics of foreign direct investment, *Hitotsubashi Journal of Economics*, Vol. 14, pp. 26-35.

Asheghian, P. (1982), Comparative efficiencies of foreign firms and local firms in Iran, *Journal of International Business Studies*, No.13 (Winter), pp.113-120.

Atri, Said and Jhun, U Jin (1990), Foreign capital inflows and domestic savings: a model with a Latent variable. *Journal of Economic Development*. Vol. 15, No.2, December, pp.25-38.

Bairam, Erkin (1987), Returns to scale, technical process and output growth in branches of industry: the case of Soviet republics, 1962-74, *Scottish Journal of Political Economy*, Vol. 34, No. 3, pp.249-266.

Bairam, Erkin (1994), *Homogeneous and Nonhomogeneous Production Functions: Theory and Applications*, Avebury: Ashgate Publishin Limited.

Baldinger, P. (1992), The birth of greater China, *The China Business Review*, May-June, pp. 13-17.

Beamish, Paul W (1987), Joint ventures in LDCs: partner selection and performance, *Management International Review*, Vol. 27, No. 1, pp.23-37.

Beamish, Paul W (1988), *Multinational Joint Ventures in Less-developing Countries*, London and New York: Routledge Press.

Beamish, Paul W (1993), Characteristics of joint ventures in the People's Republic of China, *Journal of International Marketing*, Vol. 2, pp29-48.

Beamish. Paul and Hui Y. Wang (1989), Investing in China via joint ventures, *Management International Review*, Vol. 29 No.1, pp.57-64.

Beijing Jeep Corporation (1991), *The Statistical Materials Compilation of Beijing Jeep Corporation 1984-90*.

Beijing Statistical Bureau: *Beijing Statistical Yearbook* 1988-1996, *Forty Years of Beijing*, Beijing: China Statistical Publishing House.

Bello, Walden, and Rosenfeld, Christine (1990), *Dragons in Distress: Asia's Miracle Economies in Crisis*. San Francisco: Institute for Food and Development Policy.

Bivens, Daren Kraus and Lovell, Enid Baird (1966), *Joint Ventures with Foreign Partners,* New York: the National Industrial Conference Board.

Blomstrom, Magnus (1989), *Foreign Investment and Spillovers*, Routledge Press, London and New York.

Blomstrom, Magnus and Persson, Hakan (1983), Foreign investment and spillover efficiency in an underdeveloped economy: evidence from the Mexican manufacturing industry, *World Development*, Vol.11, No.6, pp.493-501.

Bos, H. C., M. Sanders and C. Secchi (1974), *Private Foreign Investment in Developing Countries*, Dordrecht (Holland): D. Reidel Publishing Company.

Buckley, P. J. (1983), Macroeconomic versus international business approach to direct foreign investment: a comment on Professor Kojima's interpretation, *Hitotsubashi Journal of Economics*, Vol. 24, pp. 95-100.

Buckley, P. J (1985), The economic analysis of the multinational enterprise: Reading versus Japan, *Hitotsubashi Journal of Economics*, No. 26, pp. 117-124.

Buckley, Peter and Mark Casson (1985), *The Economic Theory of the Multinational Enterprise*. London: Macmillan.

Buckley, Peter and Mark Casson (1996), An economic model of international joint venture strategy, in Paul Beamish and Peter Killing (eds) *Journal of International Business Studies, Special Issue*, No.5.

Bulmer-Thomas V. (1982), *Input-Output Analysis in Developing Countries: Sources, Methods and Applications,* published by John Wiley & Sons Ltd, Chichester, New York.

Casetti, E. (1972), Generating models by the expansion method: applications to geographical research, *Geographical Analysis,* 4(1).

Casson, Mark (1985), Transaction costs and the theory of multinational enterprise, in Alan M. Rugman (ed), *New Theory of the Multinational Enterprise,* London: Croom Helm, and New York: St. Martin's Press.

Casson, M. C. and R. D. Pearce (1988), Intra-firm trade and the developing countries, *Economic Development and International Trade,* Basingstoke and London: Macmillan, pp. 132-156.

Caves, Richard (1982), *Multinational Enterprise and Economic Analysis,* New York: Cambridge University Press.

Chai, Joseph. C.H. and Leung, Chi-Keung (1985), The economic and spatial dimensions of development in China, in Joseph C.H. Chai and Chi-Keung Leung (eds) *Development and Distribution in China,* published by The University of Hong Kong.

Chai, Joseph C.H. and Yu De-lin (1986), Technology transfer and investment environment, *China's Investment Environment,* pp.312-320, Hong Kong: University of Hong Kong.

Chai, Joseph. C.H.(1994), East-West regional income gap: problems of divergent regional development in China, in Dieter Cassel/Carsten Herrmann-Phillath (eds) *The East, The West, and China's Growth: Challenge and Response,* Vol. 6, Germany: Nomos Verlagsgesellschaft, Baden-Baden.

Chen, Chung, Lawrence Chang and Yimin Zhang (1995), The role of foreign direct foreign investment in China's post-1978 economic development, *World Development* (printed in Great Britain), Vol. 23, No. 4, pp.691-703.

Chen, Edward K. Y. (1979), *Hyper-growth in Asian Economies,* London and Basingstoke: The Macmillan Press Ltd.

Chen, Edward K.Y. (1993), Foreign direct investment in East Asia, *Asian Development Review,* Vol.11, No.1, pp.24-59.

Chen Kuan, Hingchang Wang, Yuxin Zheng, Gary Jefferson and Thomas Rawski (1988), Productivity change in Chinese industry: 1953-1985, *Journal of Comparative Economics,* No.12, pp.570-591.

Chenery, Hollis B. (1961), The use of interindustry analysis in development programming, in Tibon Barna (eds) *Structural Interdependence and Economic Development*, London: Macmillan Press, 1963

Chudson, Walter A. (1985), The regulation of transfer prices by developing countries: second-best policies?, in Alan M. Rugman and Lorrain Eden (eds) *Multinationals and Transfer Pricing*, London and Sydney: Croom Helm, pp. 267-289.

Chung, V.Y.C, and Chai, J.C.H. (1993), *Taiwan's MNCs in China and ASEAN*, discussion paper, Department of Economics, The University of Queensland.

Clements, Benedict J. and Rossi, Jose W. (1991), Interindustry linkages and economic development: the case of Brazil reconsidered, *The Developing Economies*, XXIX-2, June, pp.167-186.

Cohen, Jacob and Cohen, Patricia (1983), *Applied Multiple Regression/ Correlation Analysis for the Behavioural Sciences*. Second edition. New Jersey and London: Lawrence Erlbaum Associates.

Corden, W. M. (1974), The theory of international trade, in John H. Dunning, ed., *Economics Analysis and the Multinational Enterprise*, London: Allen and Unwin, 1974, pp.184-210.

Coughlan, Anne T. and Therese Flaherty (1983), 'Measuring The International Marketing Productivity Of U.S. Semiconductor Companies', in Gautschi, David, ed., *Productivity and Efficiency in Distribution Systems*. Amsterdam: Elsevier Science Publishing Co. Pp.123-49.

Crane, George (1990). *The Political Economy of China's Special Economic Zones*. Armonk: M.E. Sharpe.

Davidson, William H (1980), The location of foreign direct investment activity: country characteristics and experience effects, *Journal of International Business Studies*, No.11, P.9-22.

Davidson, William H. (1982), *Global Strategic Management*, New York: John Wiley and Sons Press.

Dow, Sheila C. (1991), Beyond dualism, *Cambridge Journal of Economics*, Vol. 14, No. 4, pp. 143-157.

Dunning, John H. (1977), Trade, location of economic activity, and the multinational enterprise: a search for an eclectic approach, in B. Ohlin, P.O. Hesselborn and P.M. Wijkman (eds) *The International Allocation of Economic Activity*, New York: Holmes & Meier.

Dunning, John H. (1981), *International Production and the Multinational Enterprise*, London: George Allen & Unwin.

Dunning, J. H. (1985), The eclectic paradigm of international production: an update and a reply to its critics, *Mimeo* University of Reading.

Dunning, John H. (1988), The eclectic paradigm of international production: a restatement and some possible extensions, *Journal of International Business Studies*, Vol.9, No.1, pp.1-31.

Dunning, John H (1993), *Multinational Enterprises and the Global Economy*, Wokingham: Addison-Wesley Publishing Company.

Dutta, Dilip K. (1991), Socio-economic analysis of dualistic economies: a review of methodological alternatives, Journal of Interdisciplinary Economics, 3 (4), pp. 255-273)

Economist Intelligence Unit (1995), *Country Report: Taiwan*. London: EIU.

Fan, C. Cindy (1992) Regional impacts of foreign trade in China, 1984-1989, *Growth and Change: A Journal of Urban Regional Policy*, 23(2), Spring, pp.129-159.

Fujian Statistical Bureau: *Fujian Statistical yearbook* 1986-95, Beijing: China Statistical Publishing House.

Gatignon, Hubert and Anderson, Erin (1988), The multinational corporation's degree of control over foreign subsidiaries: an empirical test of a transaction cost explanation, *Journal of Law, Economics and Organization*, Vol.3-4, IV:2, pp.305-335.

Gelatt, T.A.(1989), China's New Cooperative Joint Venture Law, *Syracuse Journal of International Law and Commerce,* Vol. 15, Winter, No.2, pp.187-222.

Geroski, P. A. (1979), Review of direct foreign investment by Kiyoshi Kojima, *Economic Journal*, No. 89, pp. 162-164.

Gomes-Casseres, Benjamin (1989), Ownership structures of foreign subsidiaries: the theory and evidence, *Journal of Economic Behavior and Organization*, January, pp.1-25.

Gomes-Casseres, Benjamin (1990), Firm ownership preferences and host government restrictions: an integrated approach, *Journal of International Business Studies*, Vol.21, No.1, pp.1-22.

Goodnow, James D. and Hanz, James E. (1972), Environmental determinants of overseas market entry strategies, *Journal of International Business Studies*, No.3 (Spring), pp.33-50.

Gray, H. Peter (1993), The role of transnational corporations in international trade, in H. Peter Gray, ed. *Transnational Corporations and International Trade and Payments*, London and New York: Routledge, pp. 1-20.

Grub, Phillip Donald and Jian Hai Lin (1991), *Foreign Direct Investment in China*, New York and London: Quorum Books.

GSB (Guangdong Statistical Bureau), *Guangdong Statistical Yearbook 1985-1996,* Beijing: China Statistical Publishing House

GSB (Guangdong Statistical Bureau), *The Statistical Materials of Guangdong's Foreign Economic Relations (Guangdong Sheng Duiwai Jingji Tongji Ziliao) 1983-1992, and Annual Statistical Report of Guangdong's Industry 1992-1993.*

Haddad, Mone and Harrison, Ann (1993), Are there positive spillovers from direct foreign investment, *Journal of Development Economics*, No.42, pp.51-74, North-Holland.

Hainan Statistical Bureau: *Hainan Statistical yearbook* 1993, Beijing: China Statistical Publishing House.

Hatch, Walter, and Yamamura, Kozo (1996), *Asia in Japan's Embrace: Building a Regional Production Alliance.* Cambridge: Cambridge University Press.

Hazari, Bharat (1970), Empirical identification of key sectors in the Indian economy, *The Review of Economics and Statistics*, Vol. 52, pp. 301-305.

Hazari, Bharat and Sgro, Pasquale M. (1987), Disguised, urban unemployment and welfare in a general equilibrium model with segmented labor markets, *Journal of Regional Science*, Vol. 27, No. 3, pp. 461-475

Hebei Statistical Bureau: *Hebei Statistical Yearbook* 1989-95, and *The Forty Years of Hebei,* Beijing: China Statistical Publishing House.

Heimler, Alberto (1991), Linkages and vertical integration in the Chinese economy, *Review of Economics and Statistics*, May, pp.261-267.

Helleiner, G. K. (1973), Manufactured exports from less developed countries and multinational firms, *Economic Journal*, Vol. 9, pp. 361-374.

Hennart, Jean-Francois (1988), A transaction cost theory of equity joint ventures, *Strategic Management Journal*, Vol.9, pp.361-374.

Hennart, Jean-Francois (1991), The transaction cost theory of joint ventures: an empirical study of Japanese subsidiaries in the United States, *Management Science*, Vol.37, No.4, April, pp.483-497.

Hennart, Jean-Francois (1992), The transaction cost theory of the multinational enterprise, in Christos N. Pitelis and Roger Sugden (eds) *The Nature of the Transnational Firm*, London and New York: Routledge.

Hill, Hal (1988), *Foreign Investment and Industrialization in Indonesia*, Singapore: Oxford University Press.

Hill, Hal (1990), Foreign investment and East Asian economic development: a survey, *Asian-Pacific Economic Literature* 4 (2), pp. 21-58.

Hill, Hal and Brian Johns (1991), The role of direct foreign investment in developing East Asian countries, in Singer, Hatti and Tandon (eds) *Foreign Direct Investment* (New World Order Series: 11) pp.259-280. New Delhi: Indus Publishing Company.

Hill, Hal (1994), ASEAN economic development: an analytical survey - the state of the field', *Journal of Asian Studies* 53 (3), pp.832-66.

Hirano, M.(1993), Recent trends in investment and operations of foreign affiliates, JETRO *China Newsletter*, No. 104.

Hirschman, Albert O. (1958), *The Strategy of Economic Development*, New Haven, Yale University Press.

Hone, A. (1974): Multinational corporations and multinational buying groups: their impact on the growth of Asia's exports of manufactures: myths and realities, *World Development*, No. 2, pp. 145-149.

Hu, Michael Y. and Chen, Haiyang (1993), Foreign ownership in Chinese joint ventures: a transaction cost analysis, *Journal of Business Research*, Vol. 28, No.2, February, pp.149-160.

Hymer, Stephen H. (1960), *The International Operations of National Firms*, Cambridge Mass: MIT Press, printed in 1976).

Jansen, Karel (1995), The macroeconomic effects of direct foreign investment: the case of Thailand, *World Development*, Vol.23, No.2, pp.193-210. (England).

Jiangsu Statistical Bureau: *Jiangsu Statistical Yearbook* 1986-96, Beijing: China Statistical Publishing House.

Johnson, Harry G. (1973), Trade, investment and labor, and changing international division of production, in Helen Hughes, ed., *Prospects for Partnership, Industrialization and Trade Policies in the 1970s*, World Bank Publication.

Khan, Mohsin S. and Reinhart, Carment M (1990), Private investment and economic growth in developing countries. *Foreign Direct Investment*, chapter 9, P.243-258.

Kmenta, Jan (1967), Some properties of Alternative estimates of the Cobb-Douglas production function, *Econometrica*, Vol. 32, pp.183-1888.

Kmenta, Jan (1971), On estimation of the CES production function, *International Economic Review*, Vol. 8, pp.180-189.

Kmenta, Jan (1986), *Elements of Econometrics*, New York: Macmillan.

Kojima, Kiyoshi (1973),   A macroeconomic approach to foreign direct investment, *Hitotsubashi Journal of Economics*, Vol. 14, No. 1, pp. 1-21.

Kojima, Kiyoshi (1975), International trade and foreign investment: substitutes or complements, *Hitotsubashi Journal of Economics,* Vol. 15, No. 1.

Kojima, Kiyoshi (1978), *Direct Foreign Investment: A Japanese Model of Multinational Business Operations.* London: Croom Helm Ltd.

Kojima, Kiyoshi (1982), Macroeconomic versus international business approach to direct foreign investment, *Hitotsubashi Journal of Economics* , Vol. 23, No. 1, June, pp.1-19.

Kojima, Kiyoshi (1985), Japanese and American direct investment in Asia: a comparative analysis, *Hitotsubashi Journal of Economics* 26 (1): 1-35.

Kueh, Y.Y. (1992): Foreign investment and economic change in China, *China Quarterly,* pp.637-690.

Lardy, Nicholas R.(1980), Regional growth and income distribution in China, in Robert F. Denberger (eds) *China's Development Experience in Comparative Perspective*, pp.153-190. Cambridge, MA: Harvard University press.

Lardy, , Nicholas R. (1992), *Foreign Trade and Economic Reform in China, 1978-1990.* Cambridge: Cambridge University Press.

Lardy, Nicholas R. (1995), The role of foreign trade and investment in China's economic transformation, *The China Quarterly*, pp. 1065-1082.

Lecraw, Donald J. (1985), Some evidence on transfer pricing by multinational corporations, in Alan M. Rugman and Lorrain Eden (eds) *Multinationals and Transfer Pricing*, London and Sydney: Croom Helm, pp. 223-239.

Lee, Cheng F. and Sheng-cheng Hu (1989), *Taiwan's Foreign Investment, Exports and Financial Analysis*, (A research annual in Advances in Financial Planning and Forecasting series, supplement 1), Greenwich and Connecticut: JAI Press.

Lee, Chung H. (1984), On Japanese macroeconomic theories of direct foreign investment, *Economic Development and Cultural Change*, No. 32, pp. 713-723.

Lee, Chung H (1990), Direct foreign investment, structural adjustment, and international division of labor: a dynamic macroeconomic theory of direct foreign investment, *Hitotubash Journal of Economics*, No. 31, pp.61-72.

Lee, Jungsoo, Rana, Pradumna B. and Iwasaki, Yoshihiro (1986): Effects of foreign capital inflows on developing countries of Asia. *Asian Development Bank Economic Staff Paper*, No. 30.

Leung, Chi Kin. (1990), Locational characteristics of foreign equity joint venture investment in China, 1979-1985, *Professional Geographer*, 42, pp.403-421.

Liaoning Statistical Bureau: *Liaoning Statistical Yearbook* 1987-95, Beijing: China Statistical Publishing House.

Lim, Linda Y.C., and Pang, Eng Fong (1991), *Foreign Direct Investment and Industrialization in Malaysia, Singapore, Taiwan and Thailand*. Paris: Development Center Studies, Organization for Economic Cooperation and Development.

Little, Charles H. and Gerald A. Doeksen (1968), Measurement of linkage by the use of an input-output model, *American Journal of Agricultural Economics*, November, pp.921-934.

Lubits, R. (1971): Direct investment and capital formation, in Caves, R. E. and G. L. Reuber, *Capital Transfer and Economic Policy: Canada 1951-62*, chapter 4, Cambridge, MA: Harvard University Press.

Maddala, G.C. and Kadane, L. B. (1967), Estimation of returns to scale and the elasticity of substitution, Econometrica, Vol. 35, pp. 419-422.

Mansfield, Edwin (1982), *Microeconomics: Theory & Applications*, fourth edition, New York and London: W. W. Norton & Company.

Mason, R. Hal (1980), A comment on Professor Kojima's 'Japanese type versus American type of technology transfer, *Hitotsubashi Journal of Economics* 20 (2), pp.42-52.

McClintock, Brent (1988), Recent theories of direct foreign investment: an

McGilvary, J.W.(1977), Linkages, key industries, and development strategy, in W.W. Leontief (ed). *Structure, System and Economic Policy*, Cambridge University Press.

Ministry of Machinery Industry of China (1994), *Zhongguo Qiche Gongye Fazhan Gailan (the Survey of the Development of Automobile Industry of China) 1994.*

Ministry of Machinery Industry of China (1994), *Quanguo Qiche Gongye Fazhan Huiyi Zilao Huibian (the Proceedings of National Conference on Automobile Industry Development) 1994.*

MOFTEC (1993), *Statistical Panorama of Foreign Investment Enterprises in China 1993 (Zhongguo Waishang Touzi Qiye Tongji Daiquan),*

Beijing: published internally by the Computing Center, Ministry of Foreign Trade and Economic Cooperation.

MOFTEC (Ministry of Foreign Economic Relations and Trade of the People's Republic of China), *Almanac of China's Foreign Economic Relations and Trade*, 1984-1996. Beijing: China Social Publishing House.

MOFTEC, *Annual Statistical Report of China's Foreign Trade (Zhongguo Waimao Tongji Nianbao)* 1987-93 and *Annual Statistical Reports on Utilization of Foreign Capital (Liyong Waizi Tongji Nianbao)* 1985-1994 (unpublished statistical series).

Mundell, Robert A. (1957), International trade and factor mobility, *American Economic Review*, Vol. XLVII, No. 3, pp. 321-335.

Myers, Ramon H. (1989), Japanese imperialism in Manchuria: the South Manchurian Railway Company, 1906-1933, in Duus, Peter, Ramon H. Myers, and Mark R. Peattie (eds), *The Japanese Informal Empire in China, 1895-1937*. Princeton: Princeton University Press, pp. 101-32.

Nakagane, Katsuji (1989), Manchukuo and economic development, in Duus, Peter, Myers, Ramon H., and Peattie, Mark R.(eds), *The Japanese Informal Empire in China, 1895-1937*. Princeton: Princeton University Press. pp. 133-57.

Noorzoy, M.S. (1979): Flow of direct investment and their effects on investment in Canada, *Economic Letter*, (2), P.257-261.

Noorzoy, M.S (1980), Flow of direct investment and their effects on US. domestic investment, *Economic Letter*, (5), pp.311-317.

Nuñez, W. Peres (1990), *Foreign Direct Investment and Industrial Development in Mexico*, Paris: OECD Development Center Studies.

O'hUallachàin, Breandàn (1984), Input-output linkages and foreign direct investment in Ireland, *International Regional Science Review*, Vol.9, No.3, pp.185-200.

Ohlin, Bertil (1933), *Interregional and International Trade*, Cambridge.

Pearson, M. Margaret (1991), *Joint Ventures in the People's Republic of China: the Control of Foreign direct Investment under Socialism*. Princeton, N.J.: Princeton University Press.

Plasschaert, Sylvain R.F. (1985), Transfer pricing problems in developing countries, in Alan M Rugman and Lorrain Eden (eds) *Multinationals and Transfer Pricing,* London & Sydney: Croom Helm, pp.247-264.

Pomfret, Richard (1991), *Investing in China: Ten Years of the 'Open Door' Policy*, Harvester Wheatsheaf Press.

Purvis, Douglas D. (1972), technology, trade and factor mobility, *Economic Journal*, September, pp. 991-999.

Rana, P. B. (1985): Exports and economic growth in the Asian region. *Asian Development Bank, Economic Staff Paper*, No.25.

Rasmussen, P. (1956), *Studies in Intersectorial Relations*, Amsterdam: North-Holland.

Reynolds, Bruce L. (1987), *Reform in China: Challenges and Choice*, White Plains, New York: East Gate Books.

Robertson, D.F. and Chen, X.(1990), The new amendments to the Chinese equity joint venture law: will they stimulate foreign investment?, *Bulletin of International Bureau of Fiscal Documentation*, October, pp.484-488.

Root, Franklin (1987), *Entry Strategies for International Markets*. Lexington: Lexington Books.

Ruffin, Roy J. (1993): The role of foreign investment in the economic growth of the Asian and Pacific Region. *Asian Development Review*, Vol. 11, No. 1, pp.1-23.

Rugman, Alan (1981), *Inside the Multinationals*, New York: Colombia University Press.

Rugman, Alan (1986), New theory of multinational enterprise: an assessment of internalization theory, *Bulletin of Economic Research*, Vol.38. No.2, pp.101-118.

Santiago, Carlos E. (1987), The impact of foreign direct investment on export structure and employment generation, *World Development* (Britain), Vol.15, No.3, pp.317-328.

Schive, Chi (1982), Direct foreign investment and host country technology: a factor proportion approach in Taiwan, *Economic Essays*, No.10 (May), pp.211-227.

Schive, Chi (1990): *The Foreign Factor: The Multinational Corporation's Contribution to the Economic Modernization of the Republic of China*. Hoover Institute Press, Stanford University, California.

Schive, Chi and Majumdar, Badiul A. (1990), Direct foreign investment and linkage effects: the experience of Taiwan, *Canadian Journal of Development Studies*, Vol.11, No.2, pp.325-342.

Schultz, Siegfried (1976), Intersectoral comparison as an approach to identification of key sectors, in K. Polenske and J. V. Skolka (eds), *Advances in Input-output Analysis*, Cambridge, Mass: Ballingers.

Selwyn, M. (1993), SE Asian Chinese Head for Home, *Asian Business*, April, pp.24-29.

Shaanxi Statistical Bureau (1992 and 1993), *Shaanxi Statistical Yearbook, and the Balance Table of Inter-Provincial Capital Flows* (unpublished).

Shan, Weijian (1991), 'Environmental Risks And Joint Venture Sharing Arrangements', *Journal of International Business Studies*, No.4, pp. 555-578.

Shandong Statistical Bureau: *Shandong Statistical Yearbook* 1987-95, Beijing: China Statistical Publishing House.

Shanghai Statistical Bureau: *Shanghai Statistical Yearbook* 1987-95, and *Shanghai economic Statistical Yearbook* 1990-93, Beijing: China Statistical Publishing House.

SSB (1987), Peoples Republic of China, State Statistical Bureau (*Guojia Tongji Ju*). *Zhongguo Guding Zichan Touzi Tongji Zilao (Statistical Materials on China's Fixed-Asset Investment) 1951-85*, Beijing: the Statistical Publishing House (*Zhongguo Tongji Chuban She*).

SSB (1991), *Zhongguo Shangye Waijing Tongji Zilao 1952-1988 (Statistical Material on the Commerce and Foreign Economic Relations of China)*, Beijing: China Statistical Publishing House.

SSB (1993, 1995 and 1997), *China Foreign Economic Statistical Yearbook, 1979-1991, 1994 and 1996.* (Bilingual). Beijing: China Statistical Publishing House.

SSB (1995-1996), *China National Conditions and Power (Zhongguo Guoqing Guoli)*. Beijing: China Statistical Publishing House.

SSB (1996) *The Input-Output Table of China 1992*, Beijing: China Statistical Publishing House.

SSB (1997*), Statistical Yearbook of China's Fixed Capital Investment 1950-1995*, Beijing: China Statistical Publishing House.

SSB (1997), *The Data of The Third National Industrial Census of The People's Republic of China in 1995*, Beijing: China Statistical Publishing House.

SSB, *Statistical Yearbook of China*, 1985-1997. Beijing: China Statistical Publishing House

SSB, *Zhongguo Gongye Tongji Nianbao (Annual Statistical Report of China's Industry)*, 1987-1995, (an unpublished statistical series).

Sterner, Thomas (1990), Ownership, technology, and efficiency: an empirical study of cooperatives, multinationals, and domestic enterprises in the Mexican cement industry, *Journal of Comparative Economics*, Vol.14, pp.286-300.

Stoneman, Colin (1975), Foreign capital and economic growth. *World Development*, Vol.3, No.1, January. pp.11-22.

Stopford, John M. and Wells, Louis T. (1972), *Managing the Multinational Enterprise*, New York: Basic Books.

Sun, Haishun (1995), Direct foreign investment and regional economic development of China, *Australasian Journal of Regional Studies*, Vol. 1, No.2, pp.133-148.

Sun, Haishun (1996), Direct investment and linkage effects, *Asian Economies*, Vol. 25, No.1, pp. 5-27.

Sundrum, R.M. (1990), *Economic Growth in Theory and Practice*, London: The Macmillan Press Ltd.

Terrel, Katherine (1992): Productivity of Western and domestic capital in Polish industry, *Journal of Comparative Economics*, Vol.16, 1992, pp.494-514.

The Committee for State Regime Reforms of China (1996), *Investigation report on State-owned enterprises in 12 provinces* (in Chinese), The Chinese Age (in Chinese), p.7, July 4, 1996.

The State Council of People's Republic of China (1990), *Interim Provisions On Duration Of Chinese-Foreign Joint Ventures*, (English translation copy published in *East Asian Executive Reports*, November, 1990, pp.26-27).

The US-China Business Council (1990), *A Special Report on US Investment in China*, Washington.

The US-China Business Council (1992), The Council's Investment Initiative, *The China Business Review*, September-October, pp.6-10.

Thoburn, John T., H.M. Leung, Esther Chau and S.H. Tang (1990), *Foreign Investment in China under the Open Policy: the Experience of Hong Kong Companies*, Aldershot (England): Avebury Press.

Tianjin Statistical Bureau: *Tianjin Statistical Yearbook* 1987-1996, Beijing: China Statistical Publishing House.

Tisdell, Clem (1990a), International joint ventures and technology transfer: some economic issues, *Prometheus*, Vol.8, No.1, pp.67-78.

Tisdell, Clem (1990b), Market transaction costs and transfer pricing: consequences for the firm and for technical change, *Rivista Internazionale Di Scienze Economichee Commerciali*, Vol.37, No.3, pp.203-218.

Tisdell, Clem (1993), *Economic Development in the Context of China: Policy Issues and Analysis*, London: The Macmillan Press.

Todaro, Michael P. (1994), *Economic Development* (fifth edition), New York University, New York & London: Longman.

Tsui, Kai Yuen. (1991), China's regional inequality, 1952-1985, *Journal of Comparative Economics,* Vol.15, pp.1-21.

Tu, Jenn-Hwa (1990): *Direct foreign Investment and economic growth: a case study of Taiwan.* The Institute of Economics, Academia Sinica, Taibei, Taiwan.

UNCTAD (United Nations Conference on Trade and Development) (1996), *World Investment Report 1996: Investment, Trade and International Policy Arrangements.* United Nations: New York and Geneva, 1996.

United Nations (1992), *World Investment Report 1992,* New York.

United Nations ESCAP (1992). *Foreign Investment and Industrial Comparative Advantage in East Asia and the Pacific.* Bangkok: ST/ESCAP/1015.

Uzawa, H. (1962), Production Functions with constant elasticity of substitution, *Review of Economic Studies,* Vol. 29, pp. 291-299.

Vernon, Raymond (1966), International investment and international trade in the product cycle, *Quarterly journal of Economics,* No.80, May, pp.190-207.

Vinod, H.D. (1972), Nonhomogeneous production functions and applications to telecommunications, *Bell Journal of Economics and Management Science,* No. Vol. 3, pp. 531-543.

Walters, A.A. (1963), Production and cost functions: an econometric survey, Econometrica, Vol. 13, pp. 121-138.

Weisskoff, Richard (1977), Linkages and leakages: industrial tracking in an enclave economy, *Economic Development and Cultural Change,* Vol.25, No.4, July, pp. 607-628.

Weisskopf, T.E. (1972): The impact of foreign capital inflow on domestic savings in underdeveloped countries, *Journal of International Economics,* No.2, February, pp.25-38.

West, Guy R. (1993), *Input-Output Analysis for Practitioners: User's Guide* (unpublished computer program menu).

Williamson, Oliver (1973), Markets and Hierarchies: some elementary considerations, *American Economic Review,* No.63, pp.316-325.

Williamson, Oliver (1979), Transaction cost economics: the governance of contractual relations, *Journal of Law and Economics,* No. 22 (October), pp.233-262.

Williamson, Oliver (1981), The economics of organization: the transaction cost approach, *American Journal of Sociology*, Vol. 87, No. 3, pp.548-577.

Williamson, Oliver (1985), *The Economic Institutions of Capitalism: Firms, Markets, Relational Contracting*, New York: Free Press.

Willmore, L. (1986), The comparative performance of foreign and domestic firms in Brazil, *World Development*, Vol. 14 (April), pp.489-502.

Yamin, Mohammad (1992), A reassessment of Hymer's contribution to the theory of the transnational corporation, in Christos N. Pitelis and Roger Sugden (eds) *The Nature of the Transnational Firm*, London and New York: Routledge Press.

Yotopoulos, P.A. and Negent, J.B. (1973), A balanced growth version of the linkage hypothesis: a test, *Quarterly Journal of Economics*, No. 87, pp.157-171.

Zarembaka, P. (1970), On the empirical relevance of the CES production function, *Review of Economics and Statistics*, Vol. 52, pp. 47-53.

Zhejiang Statistical Bureau: *Zhejiang Statistical Yearbook* 1986-95, Beijing: China Statistical Publishing House.

Zweig, David (1995), 'Developmental Communities on China's Coast: The Impact of Trade, Investment, and Transnational Alliances', *Comparative Politics* 27 (3): 253-74.

Further Readings:

Alauddin, Mohamad, and Clem Tisdell (1989), Biochemical technology and Bangladeshi land productivity: Diwan and Kallianpur's analysis reapplied and critically examined, *Applied Economics*, No. 21, pp.741-760.

Alauddin, Mohamad, Dale Squires and Clem Tisdell (1993), Divergency between average and frontier production technologies: an empirical investigation for Bangladesh, *Applied Economics*, No. 25, pp. 379-388.

Arrow, K.L., Chenery, H.B., Minhas, B.S., and Slow, R.M. (1961), Capital-labor substitution and economic efficiency, *The Review of Economics and Statistics*, No.3 August, pp.225- 250.

Bruijn, E.J. and Jia, Xianfeng (1993), Transferring technology to China by means of joint ventures, *Research Technology Management*, Vol.36, Jan-Feb, pp.17-22.

Cannon, Terry.(1990), Regions, spatial inequality and regional policy, in Terry Cannon and Alan Jenkins edited *The Geography of Contemporary of China: The Impact of Deng Xiaoping's Decade,*

Debertin, David L. (1986), *Agricultural Production Economics*, New York: Macmillan Publishing Company.

Dickie, Robert B., and Layman, Thomas A. (1988), *Foreign Investment and Government Policy in the Third World: Forging Common Interests in Indonesia and Beyond.* London: Macmillan.

Falkentheim, Victor C. (1985) Spatial inequalities in China's modernization program: some political-administrative determinants, in Joseph C.H. Chai and Chi-Keung Leung (eds) *Development and urbanization in China*, pp.149-172. Hong Kong: University of Hong Kong.

Fry, Maxwell J. (1993), *Foreign Direct Investment in Southeast Asia: Different Impacts*, ISEAS Current Economic Affairs Series, ASEAN Economic Research Unit, Institute of Southeast Asian Studies. Singapore.

Jia, Liqun & Tisdell, C. (1993) Resources redistribution and regional inequality in China, *Economics Discussion Paper*, No.9318, University of Otago, New Zealand.

Johnson, Chalmers (1982), *MITI and the Japanese Miracle.* Stanford: Stanford University Press.

Kleinberg, Robert (1990), *China's 'opening' to the outside world: the experiment with foreign capitalism*, Boulder, San Francisco & Oxford: Westview Press.

Kueh, Y.Y. and Ash, Robert F. (1993), Economic integration within greater China: trade and investment flows between China, Hong Kong and Taiwan. *The China Quarterly*, pp.710-739.

Lakshmanan, T.R. and Hua, Chang-i.(1987), Regional disparities in China, *International Regional Science Reviews,* 11(1), pp.97-104.

Lo, Chor-Pang. (1989), Recent spatial restructuring in Zhujiang Delta, South China: a study of socialist regional development strategy, *Annuals of Association of American Geographers*, Vol.79, P.239-308.

Lo, Chor-Pang (1990), The geography of rural regional inequality in mainland China, *Transactions* (The Institute of British Geographers), 15, pp.466-486.

Luo, Qi and Howe, Christopher (1993), Direct investment and economic integration in the Asia Pacific: the case of Taiwanese investment in Xiamen. *The China Quarterly*, pp.747-769.

Michaely, M. (1977): Exports and economic growth: an empirical investigation, Journal of Development Economics, No. 1, March, pp. 49-53.

Miller, Arnold and Rushing, Francis W. (1990), Update China: technology transfer and trade. *Business*, January-March, pp.25-33.

Pang, Yigang (1996), Influences on foreign equity ownership level in joint ventures in China, *Journal of International Business Studies*, Vol. 27, No.1, pp.1-26.

Pannell, Cliftow W.(1987), Economic reforms and readjustment in the people's Republic of China and some geographic consequences, *Studies in Comparative International Development*, Vol. 22, No. 4, pp.54-73.

Pannell, Cliftow W (1988) Regional shifts in China's industrial output, *The Professional Geographer*, Vol. 40, No. 1, February, pp.19-32.

Phillips, David. R. and Yeh, Anthon Gar-On (1990), Foreign investment and trade: impact on spatial structure of the economy, in Terry Cannon and Alan Jenkins (eds) *The Geography of Contemporary China*, Routledge, London and New York.

Quirin, G. David (1985), Fiscal transfer pricing: accounting for reality, in Alan M Rugman and Lorrain Eden (eds) *Multinationals and Transfer Pricing,* London & Sydney: Croom Helm, pp.132-148.

Ramstetter, Eric D, ed. (1991), *Direct Foreign Investment in Asia's Developing Economies and Structural Change in the Asia-Pacific Region.* Boulder: Westview Press.

Rosenstein-Rodan, P.N. (1965), *Capital formation and economic development.* London: George Allen & Unwin Ltd.

Rothenberg, Jerome (1987), Space, interregional economic relations, and structural reform in China, *International Regional Science Reviews*, Vil.11, No. 1, pp.5-22.

Sheehey, Edmund J. (1992), Exports and growth: additional evidence, *Journal of Development Studies*, Vol. 28, No. 4, July, pp. 730-734.

Thee, Kian Wie (1991), 'The Surge of Asian NIC Investment into Indonesia', *Bulletin of Indonesian Economic Studies* 27 (3): 55-88.

Thee, Kian Wie, and Yoshihara, Kunio (1987), 'Foreign and Domestic Capital in Indonesian Industrialization', *Southeast Asian Studies* 24 (4): 327-49.

Tyler, W. (1981): Growth and export expansion in developing countries: some empirical evidence, *Journal of Development Economics*, Vol.9, No.2 August, pp.121-130.

Walker, Kenneth R.(1989), 40 years on: provincial contrasts in China's rural economic development, *the China Quarterly*, Vol. 119, pp.448-480.

Warwick, Kenneth S. (1991), Savings and investment in developing countries: sources and uses of funds, 1975-96. *Finance & Development,* June, P.36-37.

Yang, Dali (1990), Patterns of China's regional development strategy, *The China Quarterly*, June, No.22, P.231-257.

Yu, Chwo-Ming Joseph (1988a), The experience effect and foreign direct investment, *Review of World Economies*, bond (126), pp.561-580.

Yu, Chwo-Ming Joseph (1988b), New estimates of fixed investment and capital stock for Chinese state industry, *The China Quarterly*, June, Vol.114, pp.243-266.

Author Index

Subject Index